Quick Start Kubernetes

Weapons-grade learning

2024 edition

Nigel Poulton @nigelpoulton

About this edition

This is the 2024 edition. Published January 2024.

Enjoy the book and have fun with Kubernetes!

Nigel Poulton

All typos are mine. Or should that be typo's... ;-)

About the author

Nigel Poulton (@nigelpoulton)

Hi, I'm Nigel. I live in the UK and I'm a techoholic. In fact, working with technologies like the cloud, containers, and WebAssembly is *living the dream* for me!

My early career was massively influenced by a book called *Mastering Windows Server 2000* by Mark Minasi. This gave me a passion to write my own books and influence people's lives and careers the way Mark's book influenced mine. Since then, I've authored several best-selling books, including *Data Storage Networking*, *Docker Deep Dive*, and *The Kubernetes Book*. I feel immensely privileged to have reached so many people, and I genuinely appreciate all the feedback I receive.

I'm also the author of best-selling video training courses on Docker, Kubernetes, and WebAssembly. My videos are always entertaining, and occasionally laugh-out-loud funny (not my words).

When I'm not working with tech, I'm dreaming about it. When I'm not dreaming about it, I'm spending time with my family. I also like American muscle cars, coaching youth soccer, and reading sci-fi.

You can find me at all the following places, and I'm always happy to connect.

@nigelpoulton

nigelpoulton.com/books

qskbook@nigelpoulton.com

Contents

About the book

This book has two goals:

- Get you up-to-speed with Kubernetes fast
- Explain everything as clearly as possible

I've carefully chosen the most important topics and hand-crafted every chapter and example so the book is fun and engaging.

You'll love the book if you're in a hands-on role and just starting with Kubernetes. You'll also love it if you work in technical marketing, sales, management, architecture, operations, and more.

What does the book cover

The book has nine main chapters packed with theory and hands-on demos.

- **Chapter 1**: Introduces you to the concepts and clarifies important jargon
- **Chapter 2**: Explains why Kubernetes is so important
- **Chapter 3**: Gets you up-to-speed with the main components of Kubernetes
- **Chapter 4**: Shows you a couple of easy ways to get Kubernetes so you can follow along
- **Chapter 5**: Walks you through containerizing a simple app
- **Chapter 6**: Deploys the containerized app to Kubernetes
- **Chapter 7**: Demonstrates self-healing from various failures
- **Chapter 8**: Shows you how to scale an app up and down
- **Chapter 9**: Rounds everything out with a zero-downtime rolling update

What does the book cover

You'll learn ***why*** we have Kubernetes, ***what*** it is, and ***where*** it's going.

On the theory front, you'll learn about microservices, orchestration, why Kubernetes is the OS of the cloud, and the architecture of a Kubernetes cluster. On the hands-on front, you'll have the opportunity to build a cluster, containerize an app, deploy it, break it, see Kubernetes fix it, scale it, and perform an application update.

And as this is a quick start guide, you'll be up-to-speed in no time.

Will the book make you a Kubernetes expert

No, but it will start you on your journey to *becoming* an expert.

Will you know what you're talking about when you finish the book

Yes, you'll know **more than enough** to start deploying and managing simple apps on Kubernetes.

Editions

The following English language editions are available in as many Amazon territories as possible:

- Paperback
- Kindle

Ebook copies are also available from leanpub.com

The following translations are available on Amazon and Leanpub (or will be soon).

- French, German, Hindi, Italian, Portuguese, Russian, Simplified Chinese, Spanish.

Terminology and responsible language

Throughout the book, I capitalize Kubernetes API objects. Wow, we haven't even started, and I'm throwing jargon around!

Put more simply, Kubernetes *features* such as Pods and Services are spelt with a capital letter. This helps you know when I'm talking about a Kubernetes "Pod" and not a "pod" of whales.

The book also follows guidelines from the Inclusive Naming Initiative[1], which promotes responsible language.

Feedback

If you like the book and it's helped your career, share the love by recommending it to a friend and leaving a review on Amazon.

For other feedback, you can reach me at **qskbook@nigelpoulton.com**.

[1] https://inclusivenaming.org

The sample app

This is a hands-on book with a sample Node.js web app.

You can find it on GitHub at:

https://github.com/nigelpoulton/qsk-book/

Don't stress about the app and GitHub if you're not a developer. The focus of the book is Kubernetes, not the app. Plus, we explain everything as we go, and you don't have to be a GitHub expert.

The code for the app is in the **App** folder and comprises the following files.

- **app.js:** The main application file
- **bootstrap.css:** Design template for the application's web page
- **package.json:** Lists application dependencies
- **views:** Is a folder for the contents of the application's web page
- **Dockerfile:** Tells Docker how to containerize the app

If you want to download the app now, run the following command. You'll need **git** installed on your machine. It's OK if you don't have **git**, we show you how to get it and download the app later in the book.

```
$ git clone https://github.com/nigelpoulton/qsk-book.git

$ cd qsk-book
```

Finally, the app is checked at least once a year for updates and known vulnerabilities.

1: What is Kubernetes

Kubernetes is an *orchestrator* of *cloud-native microservices* applications.

That's a lot of jargon, so let's explain the following:

- Microservices
- Cloud-native
- Orchestration

Microservices

In the past, we built and deployed *monolithic applications*. That's jargon for complex applications where every feature is developed, deployed, and managed as a single large object.

Figure 1.1 shows a monolithic app with six features — web front end, authentication, middleware, logging, data store, and reporting. These are built, deployed, and managed as a single large application, meaning if you need to change any part, you need to change it all.

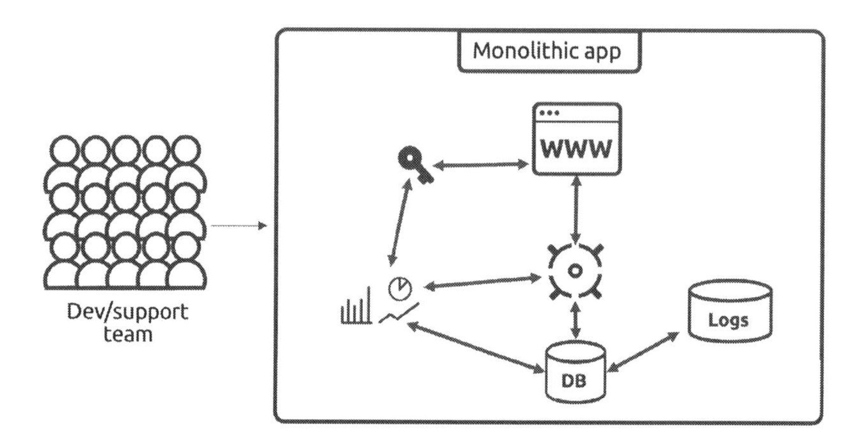

Figure 1.1

As a quick example, if you need to update the reporting feature, you need to take the entire app down and update the whole thing. This leads to high-risk updates that we have tp plan months in advance and execute over long weekends.

However, the pain of monolithic applications doesn't stop there. If you want to scale a single feature, you have to scale the whole thing.

On the flip side, *microservices applications* take the same set of features and treats each one as its own small application. Another word for "small" is "micro", and another word for "application" is "service". Hence, the term *microservice*.

If you look closely at Figure 1.2, you'll see it's the exact same application as Figure 1.1. The difference is that each feature is developed independently, each is deployed independently, and each can be updated and scaled independently. But they work together to create the exact same *application experience.*

The most common pattern is developing and deploying each microservice as its own container. This way, if you need to scale the reporting service, you just add more reporting containers. If you need to update the reporting service, deploy a new reporting container and delete the old one.

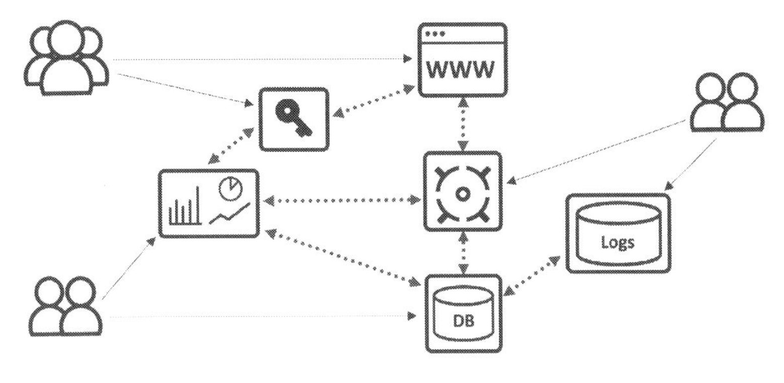

Figure 1.2

Microservices are loosely coupled by design, and each one exposes an API that others use to consume it. These are fundamental to the ability to change one without affecting others.

The following car analogy might help if you're new to the concept of APIs.

Cars come in all shapes and sizes — sports cars, SUVs, trucks, petrol, diesel, electric, hybrid, hydrogen fuel cell, etc. However, these differences are hidden from drivers behind a standard set of controls, including a steering wheel and foot pedals. In this model, the steering wheel and foot pedals are the car's API – how we *consume its capabilities*. This means a driver can get into any car in the world and be able to drive it. For example, I learned to drive in a front-wheel-drive petrol-engine car with the steering wheel on the right. However, I can step into an all-wheel drive electric car with the steering wheel on the left and be able to drive it without having to learn any new skills.

Well, it's the same with microservices applications. As long as you don't change the API for a microservice, you can patch or update it without impacting consumers.

As well as the ability to update and scale individual features, microservices lends itself to smaller and more agile development teams that can deliver faster. It's common to apply the *two-pizza team rule* that states *if you can't feed a development team on two pizzas, the team is too big.*

However, microservices introduce their own challenges. They can become very complex, with lots of moving parts, owned and managed by different teams. This needs good processes and good communication.

Finally, both of these — monolithic and microservices — are called *design patterns.* The microservices design pattern is the most common pattern in the current cloud era.

Cloud-native

This is easy, as we've covered some of it already.

A *cloud-native* app must:

- Self-heal
- Scale on demand
- Support rolling updates

Let's unpick some of that jargon.

Scaling on demand is the ability for an application and associated infrastructure to grow and shrink automatically to meet requirements. For example, an online retail app might need to scale up infrastructure and application resources during holiday periods. If configured correctly, Kubernetes can automatically scale applications and infrastructure according to demand.

Not only does this help businesses react more quickly to unexpected changes, it also reduces infrastructure costs when scaling down.

Kubernetes can also *self-heal* applications. You tell Kubernetes what an app should look like, such as how many instances of each microservice. Kubernetes records this as *desired state* and watches the app to make sure it always matches *desired state.* When things change, such as a microservice failing, Kubernetes notices this and spins up a replacement. We call this *self-healing* or *resiliency*.

Rolling updates is the ability to update parts of an application without taking it offline and impacting consumers. It's a game-changer in today's always-on world, and we'll see it in action later.

One final point. *Cloud-native* has almost nothing to do with the public cloud. For example, deploying a monolithic application to the cloud does **not** make it cloud-native. Whereas a microservices application that self-heals, automatically scales and does rolling updates, deployed to your on-premises datacenter **is** cloud native. We call applications *cloud-native* because they possess the attributes we associate with public clouds — resilient, elastic, always on...

In summary, cloud-native apps are resilient, automatically scale, and can be updated without downtime.

Orchestration

Orchestration can be a difficult concept to understand. The following sports analogy should help.

A football (soccer) team is a group of individual players. Each has a different set of skills and attributes, and each has a different role to play when the game starts.

Figure 1.3 shows an unorganised football team without a game plan.

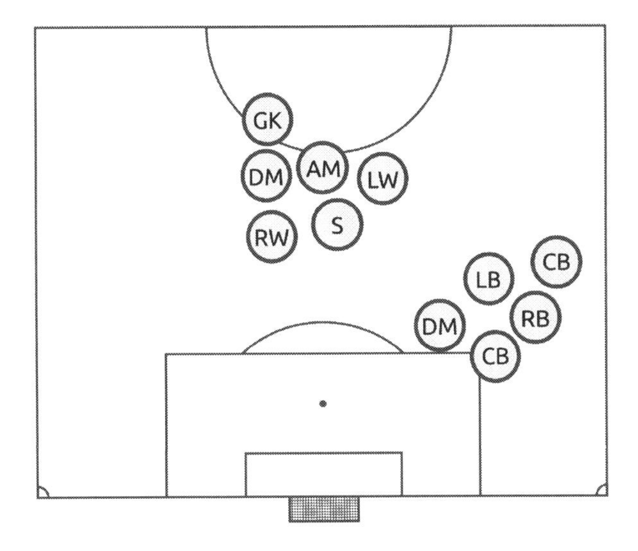

Figure 1.3

The coach comes along with the game plan and assigns each player a position and a job. She also dictates the tempo of the game, such as when to press, how high to press, and how quickly to get the ball into the opposition's box.

In short, the coach takes the chaos from Figure 1.3 and imposes the order in

Figure 1.4. She also reacts to real-time events, such as injuries and tactical changes depending on the score.

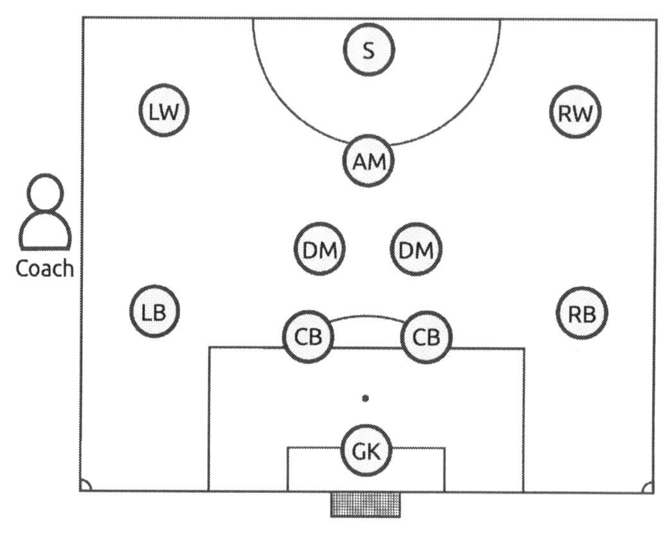

Figure 1.4

Well... cloud-native microservices applications are a lot like sports teams.

Each cloud-native app is made of individual microservices that do different things. Some serve web requests, some authenticate, some do logging, some persist data, some generate reports, etc. And, just like a sports team, they need something to organise them into a useful app.

Enter Kubernetes.

Kubernetes takes a mess of independent microservices and organises them into meaningful apps, as shown in Figure 1.5. It also responds to real-time events by self-healing, scaling and more.

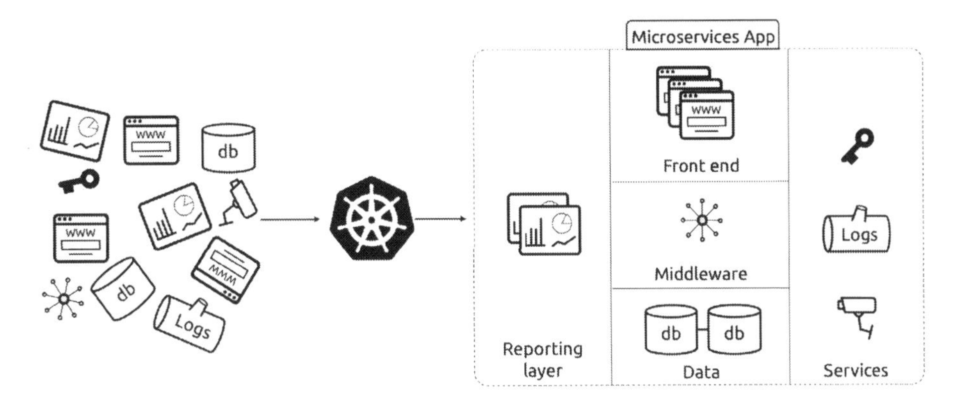

Figure 1.5

In summary, Kubernetes is an *application orchestrator* that brings together different microservices and organises them into a useful application. It also provides and manages cloud-native features such as scaling, self-healing, and updates.

Other useful Kubernetes things to know

The name *Kubernetes* comes from the Greek word meaning *helmsman*. This is a nautical term for the person that steers a ship.

The wheel of a ship is called the *helm* and is where the Kubernetes logo comes from.

Figure 1.6. Kubernetes logo.

If you look closely, you'll see the logo has seven spokes instead of the usual 6 or 8. This is because the original Kubernetes developers had worked on the Google **Borg** project and wanted to name Kubernetes *"Seven of Nine"* after the famous

Borg drone rescued by the USS Voyager on stardate 25479. Copyright laws prevented this, so the founders gave the logo *seven* spokes in a subtle reference to "**Seven** of Nine".

You'll also see *Kubernetes* shortened to *K8s*. The *8* replaces the eight letters between the leading "K" and the trailing "s". It's usually pronounced "kates".

Chapter summary

At the top of the chapter, we said that *Kubernetes is an orchestrator of cloud-native microservices applications.*

Now that we've explained the jargon, you know this means *"Kubernetes runs and manages applications comprised of small specialised parts that can self-heal, scale, and be updated independently without downtime."* Those specialised parts are called *microservices* and each one is usually deployed in its own container.

However, that's still a lot, and you don't need to understand everything yet. We'll continue explaining things, and we'll get hands-on with lots of examples that will help.

2: Why Kubernetes is so important

As the title suggests, this chapter explains *why* we need Kubernetes. We'll cover two main areas:

- Why the cloud providers need Kubernetes
- Why the user community needs Kubernetes

Both are important, and both are part of why Kubernetes will be with us for a long time. Some of the things we'll discuss will help you avoid pitfalls when getting started.

Why the cloud providers need Kubernetes

It all starts with Amazon Web Services (AWS).

Prior to 2006, the big tech companies were making easy money selling servers, network switches, storage arrays, and licenses for monolithic apps. Then, from way out in the left field, Amazon launched AWS and turned the world upside down. It was the birth of modern cloud computing.

At first, the big tech companies didn't seem to care. Remember, they were too busy making money doing the same things they'd been doing for decades. However, as soon as AWS started stealing customers, the rest of the industry needed a response.

The first response was to debunk AWS by claiming there was no such thing as the cloud. When that didn't work, they re-invented themselves as *cloud companies* and started competing against AWS.

An early attempt was OpenStack[2]. To keep a long story short, OpenStack was a community project building an open-source alternative to AWS. It was a great

[2]https://www.openstack.org/

project, and lots of amazing people contributed. In fact, OpenStack still exists, it just never threatened AWS.

While all of this was happening, Google was running most of its services at massive scale on Linux containers. Things like Search and Gmail were churning through billions of containers per week, and scheduling all of these were a couple of in-house tools called *Borg* and *Omega*.

Fast-forward a few years, and some of the *Borg* and *Omega* engineers at Google took what they'd learned and built a new container platform called *Kubernetes*. They open-sourced the project and donated it to the Cloud Native Computing Foundation (CNCF) in 2014 as the first-ever CNCF project.

Now then, Kubernetes is not an open-source version of Borg or Omega. It's a new project, built from scratch. Its only connection to *Borg* and *Omega* is that its initial developers worked on those projects at Google. However, Kubernetes is ~10 years old and has gone off in its own direction since then.

When Google open-sourced Kubernetes in 2014, Docker was taking the world by storm. This caused most people to see Kubernetes as a way to manage the explosive growth of containers. And while that's true, it's only half the story. Kubernetes is also excellent at *abstracting* and *commoditizing* cloud and server infrastructure.

Abstracting and *commoditizing* infrastructure makes Kubernetes a lot like traditional operating systems such as Linux and Windows. For example, Linux and Windows make it so we don't have to care if our traditional apps run on Cisco, Dell, HPE, or XYZ servers. Kubernetes does the same by making it so we don't have to care if our cloud-native apps run on AWS, Azure, Civo Cloud, or servers in our datacenter. Figure 2.1 shows a cloud-native app that can run on any of the four platforms. This is why you'll hear Kubernetes referred to as the *OS of the cloud*.

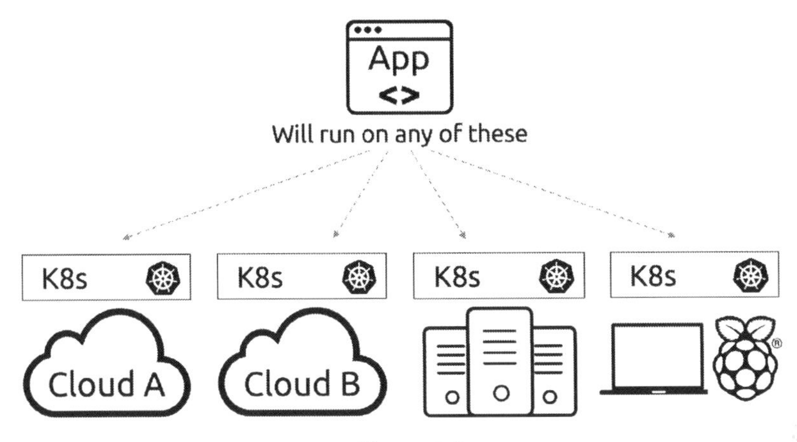

Figure 2.1

Abstracting cloud infrastructure meant competing cloud vendors could use Kubernetes to wipe out some of the value of AWS — if they can get users to build applications to run on Kubernetes, it shouldn't matter which cloud they run on. This is why cloud providers place Kubernetes front-and-center in their offerings.

Why the user community needs Kubernetes

The user community needs vendor-neutral platforms that provide flexibility and have a strong future. As things stand, Kubernetes fits the bill.

Kubernetes is an open-source project hosted and maintained by the Cloud Native Computing Foundation (CNCF). The CNCF is a Linux Foundation project with a goal of creating a vendor-neutral cloud. Of course, some vendors have more influence than others, but Kubernetes has remained vendor-neutral so far.

As the *OS of the cloud,* Kubernetes gives users great flexibility and helps avoid cloud lock-in.

Most of the major cloud vendors contribute to the *upstream* Kubernetes project and use this as the basis of their own *hosted Kubernetes services.* This creates a strong future for Kubernetes.

Figure 2.2 shows the upstream open-source Kubernetes project and how it relates to vendor implementations. The upstream project is where the new features and new development happen. Cloud vendors take this and use it to build their own cloud platforms and services. The diagram is very high-level and only for illustration purposes.

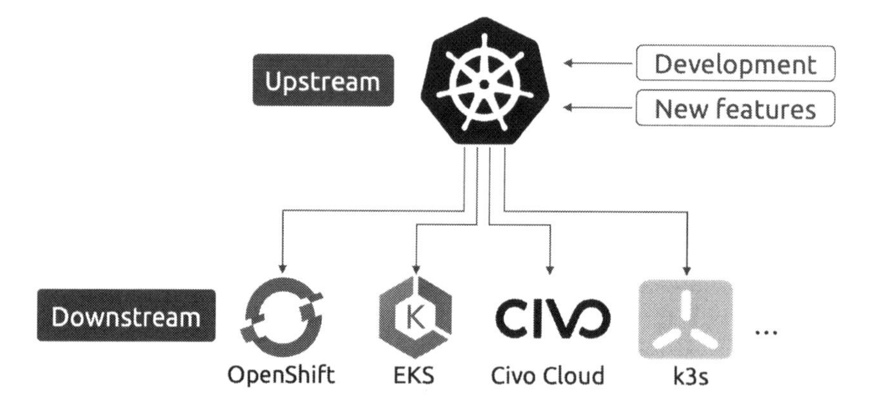

Figure 2.2

As things stand, Kubernetes is vendor-neutral, enables cloud flexibility, and has a strong future.

Chapter Summary

In this chapter, you learned that the major cloud providers are heavily invested in the success of Kubernetes. This creates a strong future for Kubernetes and makes it a safe platform for users and companies to build on. Kubernetes also abstracts underlying infrastructure the same way operating systems like Linux and Windows do. This is why it's referred to as the *OS of the cloud.*

3: Kubernetes architecture

We've already said Kubernetes sits between applications and infrastructure and acts like the *OS of the cloud.* Figure 3.1 shows applications running on Kubernetes, which, in turn, runs on infrastructure.

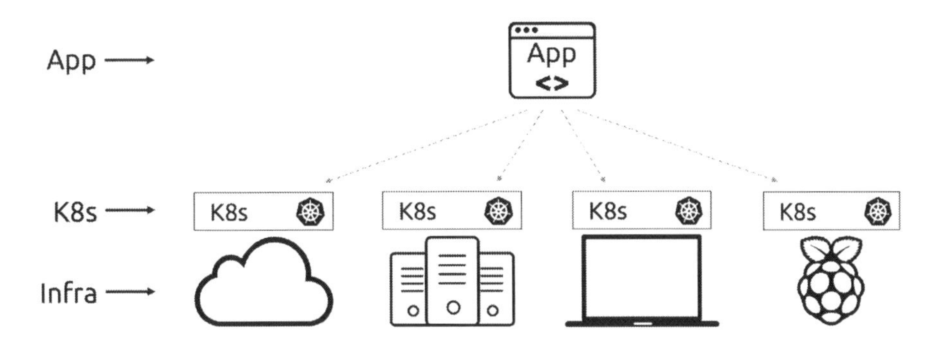

Figure 3.1

The diagram shows four Kubernetes installations running on four different platforms. All the application can see is four Kubernetes clusters, it can't see the infrastructure below. This makes it easier to migrate applications from one Kubernetes installation to another.

We call a Kubernetes installation a *Kubernetes cluster*.

There are a couple of things worth clarifying about Figure 3.1.

Firstly, it's unusual for a single cluster to span multiple infrastructures. For example, you aren't likely to see a cluster spanning multiple clouds. Likewise, you're unlikely to see one spanning on-prem and a public cloud. This is mainly because Kubernetes needs reliable low-latency networks between nodes.

Secondly, although Kubernetes can run on many platforms, containers have stricter requirements. You'll see this later, but Windows containers need Windows cluster nodes, and Linux containers need Linux cluster nodes. The same applies to CPU architectures — a container built for AMD/x86 won't run on cluster nodes with ARM CPUs.

Control plane nodes and worker nodes

A *Kubernetes cluster* is one or more machines with Kubernetes installed. These *machines* can be physical servers, virtual machines (VM), cloud instances, your laptop, Raspberry Pis, and more. Installing Kubernetes on them and connecting them together creates a *cluster*. We deploy applications to clusters.

We refer to machines in a Kubernetes cluster as *nodes*, and there are two types:

- Control plane nodes
- Worker nodes

Figure 3.2 shows a six-node Kubernetes cluster with three control plane nodes and three worker nodes. It's usually a good practice to run apps on *worker nodes* and reserve the *control plane nodes* for Kubernetes system services.

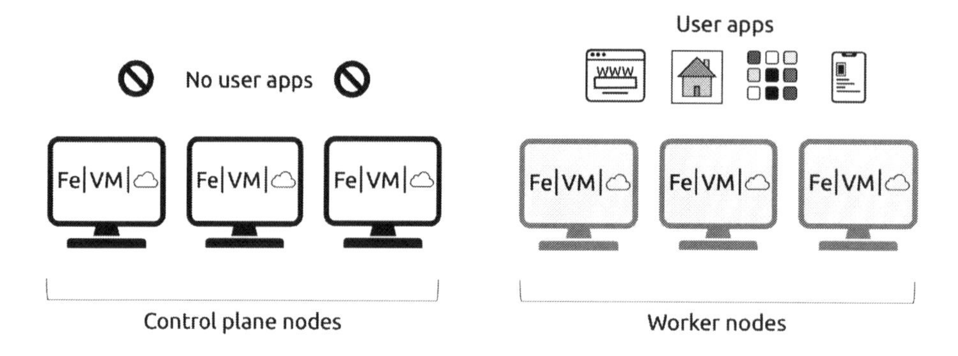

Figure 3.2

Control plane nodes

Control plane nodes run the internal Kubernetes system services. These include the API server, the scheduler, the cluster store, and more. Collectively, we refer to them as the *control plane*. Each control plane node runs every control plane service.

With this in mind, it's a good practice to have multiple control plane nodes for high availability (HA). This way, if one of them fails, the cluster keeps running.

In the real world, it's common for production clusters to have three or five control plane nodes and to spread them across failure domains. Do **not** put them all in the same rack under the same leaky aircon unit on the same glitchy power supply.

Figure 3.3 shows a highly-available control plane with three nodes. Each one is in a separate failure domain with separate network and power infrastructures etc.

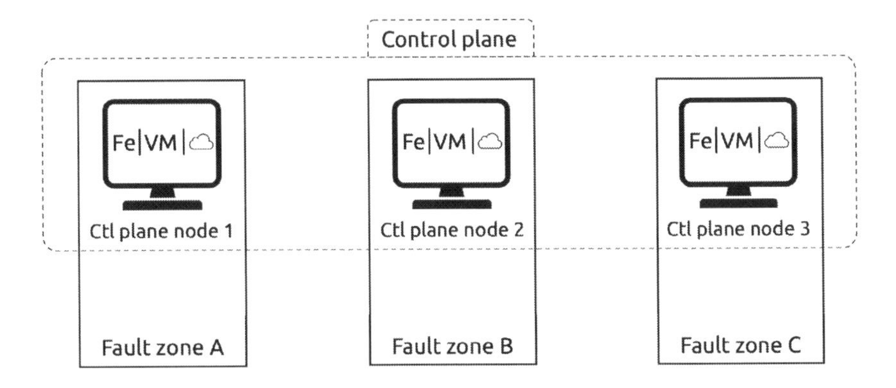

Figure 3.3. Control plane HA.

As previously mentioned, the control plane is made up of the following services:

- API Server
- Scheduler
- Store
- Cloud controller
- More…

The *API Server* is the **only** part of a Kubernetes cluster you interact directly with. For example, you send commands to deploy, manage, and update apps to the API server. You even send queries about the state of applications to the API server. In this case, the API server queries the cluster store and sends the response.

The *Scheduler* chooses which worker nodes to run applications on.

The *Store* is where application and cluster state are kept.

The *Cloud controller* integrates Kubernetes with cloud services such as storage and load balancers.

There are more control plane services, but those are the important ones.

Worker nodes

Worker nodes are where user applications run and can be Linux or Windows. Linux apps run on Linux nodes, and Windows apps run on Windows nodes. Fortunately, a single cluster can have a mix of node types.

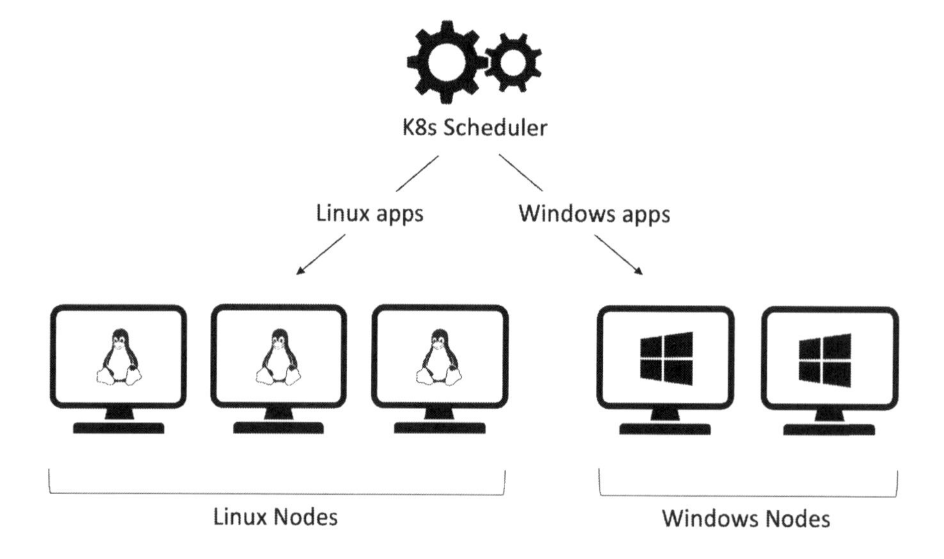

Figure 3.4. Cluster with Linux and Windows worker nodes.

Worker nodes run two essential services:

- Kubelet
- Container runtime

The *kubelet* is the main Kubernetes agent. It joins workers to the cluster and communicates with the control plane. For example, it watches the API server for new work tasks, and it sends status reports back to the API server.

The *container runtime* manages container lifecycle events such as creating, starting, stopping, and deleting.

You should know that Kubernetes used to use Docker as its *container runtime.* However, in 2016, Kubernetes introduced the container runtime interface (CRI) that allows you to pick and choose different runtimes. Since then, **containerd** (pronounced "container dee") has replaced Docker as the default container runtime in most Kubernetes environments. It's a stripped-down version of Docker and fully supports container images created by Docker. Lots of other runtimes exist, but they're beyond the scope of a quick start book. See *The Kubernetes Book* for more detail.

Hosted Kubernetes

Hosted Kubernetes is a consumption model where a cloud provider rents you a production-grade Kubernetes cluster. If you're following along on the Civo Cloud, this is a hosted Kubernetes.

The cloud provider builds the cluster, owns the control plane, and is responsible for all of the following:

- Control plane performance
- Control plane availability
- Control plane updates

You're usually responsible for:

- Worker nodes
- User applications
- Paying the bill

Figure 3.5 shows the basic architecture and division of responsibility for hosted Kubernetes platforms.

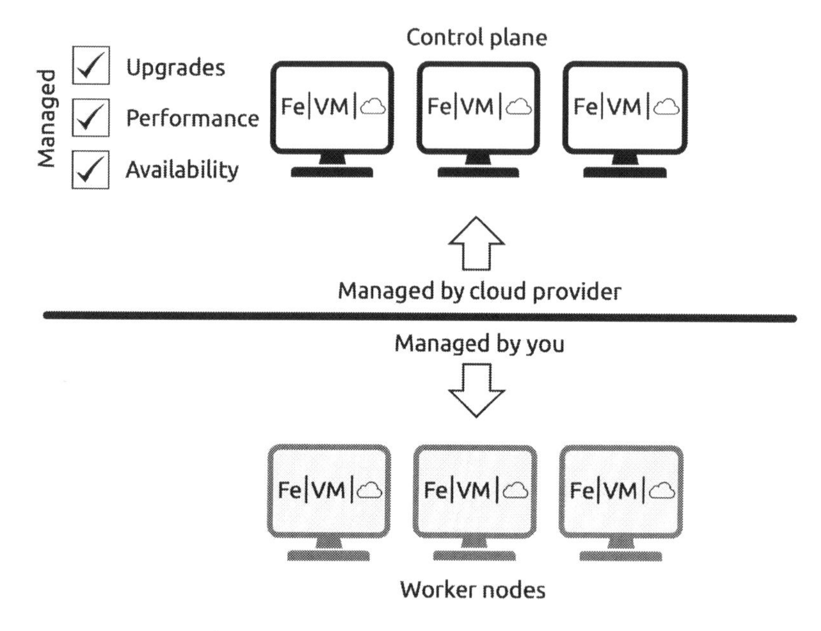

Figure 3.5. Hosted Kubernetes architecture

Most of the cloud providers have hosted Kubernetes services. Some of the more popular ones include:

- AWS: Elastic Kubernetes Service (EKS)
- Azure: Azure Kubernetes Service (AKS)
- Civo: Civo Cloud Kubernetes
- DO: Digital Ocean Kubernetes Service (DOKS)
- GCP: Google Kubernetes Engine (GKE)
- Linode: Linode Kubernetes Engine (LKE)

Others exist, and not all services are equal. For example, Civo Cloud Kubernetes is clean and easy to use. However, it may lack some of the integrations and configuration options offered by others. You should try a few before deciding which is best for you.

Managing Kubernetes with the kubectl command line tool

Most day-to-day management of Kubernetes is done using the **kubectl** command line tool. There are lots of ways to pronounce it, and all ways are acceptable. I pronounce it *"kube see tee ell"*.

Management tasks include things such as deploying and managing applications, checking the health of the cluster and applications, and performing updates.

You can get kubectl for Linux, macOS, Windows, and more, and you'll see how to install it in the next chapter.

The following example command lists all nodes in a cluster. You'll run plenty of commands in the hands-on sections later.

```
$ kubectl get nodes
NAME            STATUS    ROLES                       AGE     VERSION
qsk-server-0    Ready     control-plane,etcd,         15s     v1.29.0
qsk-agent-2     Ready     <none>                      15s     v1.29.0
qsk-agent-0     Ready     <none>                      13s     v1.29.0
qsk-agent-1     Ready     <none>                      10s     v1.29.0
```

Chapter summary

In this chapter, you learned that Kubernetes clusters are made up of *control plane nodes* and *worker nodes*. These can run almost anywhere, including bare metal servers, virtual machines, and in the cloud. Control plane nodes run the backend services that keep the cluster running, whereas worker nodes run business applications.

Most cloud platforms offer a *hosted Kubernetes service*. These are an easy way to get a *production-grade* cluster where the cloud provider manages performance, availability, and updates. You manage the worker nodes and pay the bill.

You also learned that *kubectl* is the Kubernetes command line tool.

4: Getting Kubernetes

The goal of this chapter is to get you a lab environment so you can follow along with the examples in the book.

To do this, you'll need all of the following:

- Docker
- kubectl
- A Kubernetes cluster
- The sample app

Installing Docker Desktop will get you Docker and kubectl. You can also enable Docker Desktop's built-in single-node Kubernetes cluster. It's the easiest way to get everything, but you won't be able to use the single-node cluster for every example.

If you want to follow every example, you should install Docker Desktop and then create a multi-node Kubernetes cluster in the cloud. We'll show you how to do this on the Civo Cloud.

There are lots of other ways to get Docker, kubectl, and a Kubernetes cluster. However, the ways we'll show you are the easiest and cheapest.

The chapter is divided as follows:

- Get Docker and kubectl with Docker Desktop
- Get a single-node Kubernetes cluster with Docker Desktop
- Get a multi-node Kubernetes cluster with Civo Cloud
- Get the sample app

Readers get $500 of free credit on Civo Cloud. This is more than enough to complete all the examples in the book, and the credit lasts for three months from the date you register.

Get Docker and kubectl with Docker Desktop

You can get Docker Desktop on Windows, macOS, and Linux. It's easy to install, and you get Docker and kubectl.

> **Note:** Docker Desktop is free for personal use, but you'll have to pay a license fee if you use it for work and your company has more than 250 employees or does more than $10M in annual revenue.

Type "download docker desktop" into your favorite search engine and follow the links to download the installer for your system. After that, it's a *next next next* installer that requires admin privileges. Windows users should install **WSL 2** if prompted.

After the installation completes, you may need to start the app manually.

When it's running, Mac users get a Docker whale in the menu bar, whereas Windows users get it in the system tray at the bottom.

Run the following commands to check you have Docker and kubectl.

```
$ docker --version
Docker version 24.0.6, build ed223bc

$ kubectl version --client
Client Version: v1.28.2
```

Congratulations, you have Docker and kubectl.

Get a single-node Kubernetes cluster with Docker Desktop

Docker Desktop ships with a built-in single-node Kubernetes cluster. All you have to do is enable it, and you'll be able to use it for most of the examples in the book. However, You won't be able to use it for the examples that require multiple nodes.

Skip to the next section if you plan to use a multi-node cluster on the Civo Cloud.

Complete the following steps to enable the Docker Desktop Kubernetes cluster.

1. Click the whale icon
2. Choose **Settings**
3. Open the **Kubernetes** tab
4. Check the **Enable Kubernetes** checkbox
5. Click **Apply & restart**

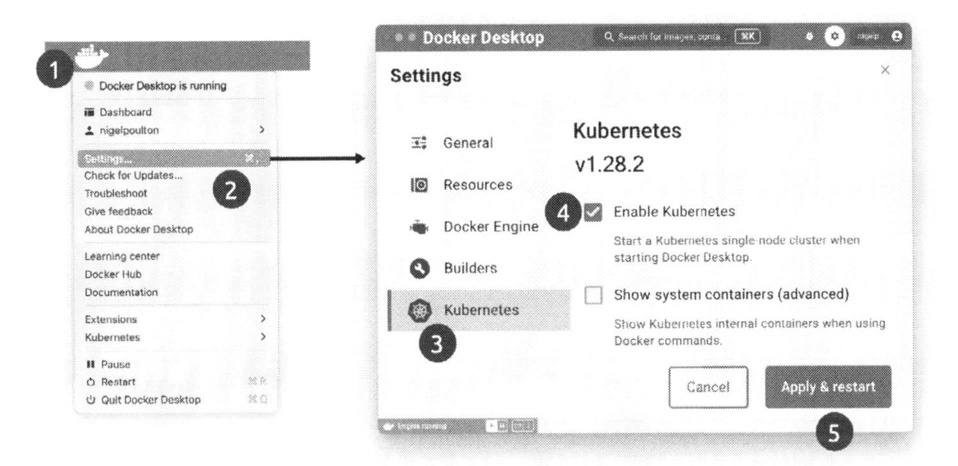

Figure 4.1

It can take a couple of minutes for the cluster to install.

Windows users need to switch Docker Desktop into ***Linux containers*** mode to follow along with the examples. To do this, right-click the Docker whale in the system tray and choose **Switch to Linux containers**. This allows Windows machines to run Linux containers.

Run the following command to make sure the cluster is up and running. If the command times out, it's probably because the cluster hasn't finished installing.

```
$ kubectl get nodes
NAME             STATUS   ROLES           AGE   VERSION
docker-desktop   Ready    control-plane   15h   v1.28.2
```

The cluster only has a single node that hosts all control plane features and will run all user apps.

If you plan on using this cluster, you have everything you need and can skip to the **Get the sample app** section.

Get a multi-node Kubernetes cluster with Civo Cloud

Most clouds have a hosted Kubernetes service, and you can follow along on any. However, the book uses Civo Cloud Kubernetes because it's easy to use, and readers get $500 of free credit using the **civo.com/nigel** sign-up link. The free credit is more than enough to complete the examples, it lasts for three months following sign-up, and the link will work until at least 2025.

What you get with Civo Cloud Kubernetes

Civo Cloud Kubernetes is a hosted Kubernetes service. As such:

- It's easy to setup
- It's multi-node
- It's production-grade
- The control plane is managed by Civo and hidden from you
- It offers advanced integrations with cloud services such as storage and load balancers

What you don't get with Civo Cloud Kubernetes

Civo Cloud Kubernetes doesn't give you Docker or kubectl. You'll need these to follow along, and the easiest way to get them is to install Docker Desktop.

Get a Civo Cloud Kubernetes cluster

Point your browser to **civo.com/nigel** and sign up for an account. It's a simple process, and you'll automatically get $500 of free credit that lasts for three months. You have to provide billing details, but the free credit will be more than enough to complete all the examples in the book.

Once you're set up, log in to the Civo Dashboard, click **Kubernetes** from the left navigation bar and choose **Create new Cluster**.

Give your cluster the following settings:

- **Name:** qsk
- **How many nodes:** 3
- **Select a size:** Small
- **Network:** default
- **Firewall:** default
- Expand the **Show advanced options** section and choose a **k3s** cluster at version 1.28 or higher.

Leave all other options as default and click **Create Cluster**.

It'll take a couple of minutes for the cluster to build.

When it's ready, the Civo dashboard will show basic info about your highly-available multi-node cluster.

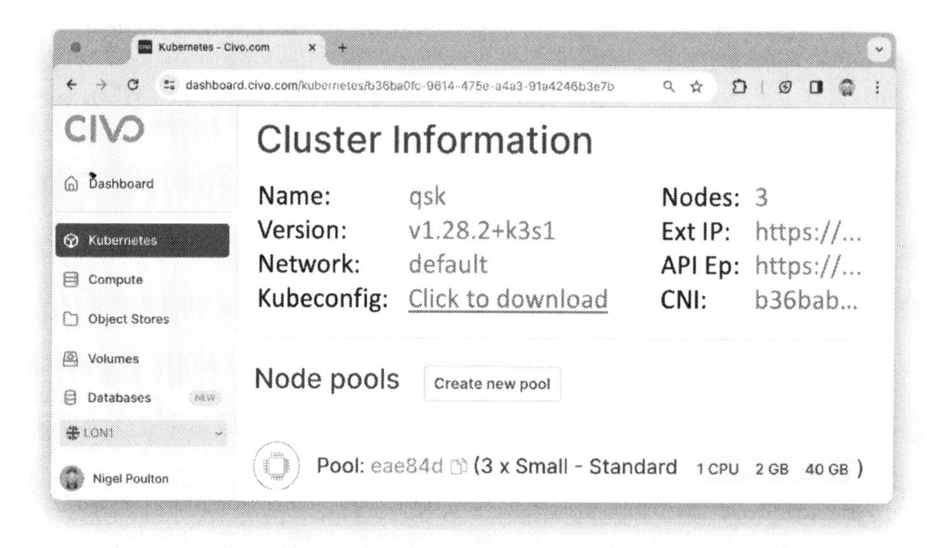

Figure 4.2. Civo Cloud Kubernetes cluster.

Your cluster is ready, but you need to configure kubectl.

Configure kubectl

kubectl uses a config file to know which cluster to manage and which credentials to use.

The file is called *config* and is located in the following hidden directories on Mac and Windows. We usually call it the *kubeconfig file*.

- Windows: C:\Users\<username>\.kube\config
- macOS: /Users/<username>/.kube/config

The easiest way to configure kubectl for your new Civo Cloud cluster is to:

1. Rename your existing *kubeconfig* file
2. Download and use the *kubeconfig* file from the Civo dashboard

For the following to work, you'll need to configure your computer to show hidden folders. On macOS, open a **Finder** window and type **Command** + **Shift** + **period**. On Windows 10 or 11, type "folder" into the Windows search bar and select the **File Explorer Options** result. Select the **View** tab and click the **Show hidden files, folders, and drives** button. Remember to click the **Apply** button.

Navigate to your system's hidden **.kube** directory and rename the existing "**config**" file. Feel free to rename it to anything you like, and it's OK if the file doesn't exist. If the directory doesn't exist, create it and be sure to include the leading dot (.) to make it a hidden directory.

Select your cluster in the Civo dashboard and download the *Kubeconfig* file from the **Cluster Information** section, as shown in Figure 4.3.

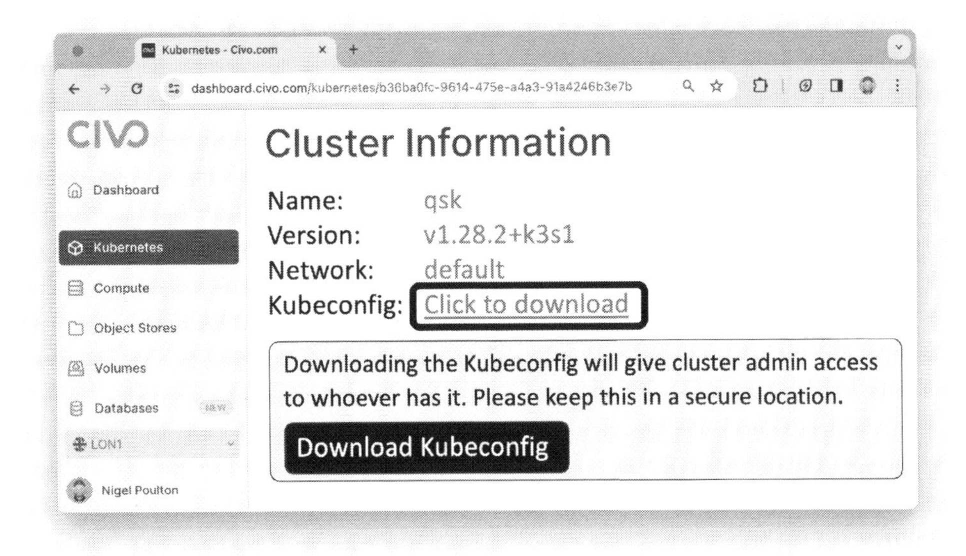

Figure 4.3. Download kubeconfig file.

Locate the downloaded file, copy it to the hidden `./kube` folder in your home directory, and rename it to "**config**".

Once you've done this, run the following command to verify that kubectl is talking to your Civo Cloud cluster.

```
$ kubectl get nodes
NAME               STATUS   ROLES    AGE   VERSION
k3s-qsk...d5idm    Ready    <none>   5m    v1.28.2+k3s1
k3s-qsk...jhnpn    Ready    <none>   5m    v1.28.2+k3s1
k3s-qsk...um2r2    Ready    <none>   5m    v1.28.2+k3s1
```

The output shows a cluster with three worker nodes. You know it's your Civo Kubernetes cluster because the node names include the name of your cluster prefixed with "k3s" (`k3s-qsk...`). You'll only see worker nodes as the control plane nodes are managed by Civo and not displayed.

At this point, your Civo Cloud Kubernetes cluster is up and running and you can use it to follow all the examples in the book.

Remember to delete it when you no longer need it. Forgetting to do this will waste energy and may incur unwanted costs.

Get the sample app

The sample app and config files are located on GitHub. The easiest way to get them is to *clone* the repo with the **git** command line tool.

Don't worry if you don't know how to use Git, nobody does ;-)

If you don't already have it, Mac users can use Homebrew to install the git CLI. Windows users can use chocolatey or other package managers. For other methods, search the internet for how to install the *git CLI* and follow the instructions for your system.

Once it's installed, run the following command to download the sample app. It will create a new directory, called `qsk-book`, with all the files you need to follow along.

```
$ git clone https://github.com/nigelpoulton/qsk-book.git
Cloning into 'qsk-book'...
remote: Enumerating objects: 134, done.
remote: Counting objects: 100% (30/30), done.
remote: Compressing objects: 100% (25/25), done.
remote: Total 134 (delta 7), reused 21 (delta 3), pack-reused 104
Receiving objects: 100% (134/134), 74.92 KiB | 235.00 KiB/s, done.
Resolving deltas: 100% (53/53), done.
```

Change into the **qsk-book** directory and list the files.

```
$ cd qsk-book

$ ls -l
drwxr-xr-x  7 nigelpoulton  staff  224  App
drwxr-xr-x  8 nigelpoulton  staff  256  Appv1.1
-rw-r--r--  1 nigelpoulton  staff  390  deploy.yml
-rw-r--r--  1 nigelpoulton  staff  225  pod.yml
-rw-r--r--  1 nigelpoulton  staff  929  readme.md
-rw-r--r--  1 nigelpoulton  staff  509  rolling-update.yml
-rw-r--r--  1 nigelpoulton  staff  217  svc.yml
```

You're ready to follow along with the demos.

Chapter summary

Docker Desktop is a great way to get a single-node Kubernetes cluster. It's free to use for personal projects and personal learning, and it automatically installs and configures *kubectl*. It's not intended for production use, but you can use it to follow along with most of the examples.

Civo Cloud Kubernetes is a simple-to-use hosted Kubernetes service. Civo manages the control plane and lets you size and spec as many worker nodes as you need. Readers get $500 of free credit that lasts for three months and is more than enough to complete the examples in the book. However, remember to delete your cluster when you're finished so you don't incur unexpected costs and waste energy.

There are lots of other ways to get Kubernetes, but the ways we've shown here are enough to get you started.

5: Containerizing an app

In this chapter, you'll build an application into a container image. The process is called *containerization* and the resulting app is called a *containerized app.*

You'll use Docker to containerize the app, and the steps are not specific to Kubernetes. In fact, you won't use Kubernetes in this chapter. However, it's a vital part of a typical Kubernetes workflow, and you'll deploy the containerized app to Kubernetes in the following chapters.

> **Docker and Kubernetes:** Let's clear up the confusion about Kubernetes supposedly dropping support for Docker. Kubernetes stopped using Docker as a container runtime in version 1.24. This means Kubernetes 1.24 and later do not use Docker to start and stop containers. However, apps containerized by Docker still work on Kubernetes and probably make up the majority of apps running on Kubernetes. This is because Kubernetes and Docker both implement Open Container Initiative (OCI) standards.

You can skip this chapter if you're already familiar with containerizing apps. There's a pre-created image on Docker Hub you can use.

The workflow you'll follow is shown in Figure 5.1. We'll briefly touch on step 1, but the focus will be on steps 2 and 3. Future chapters will cover step 4.

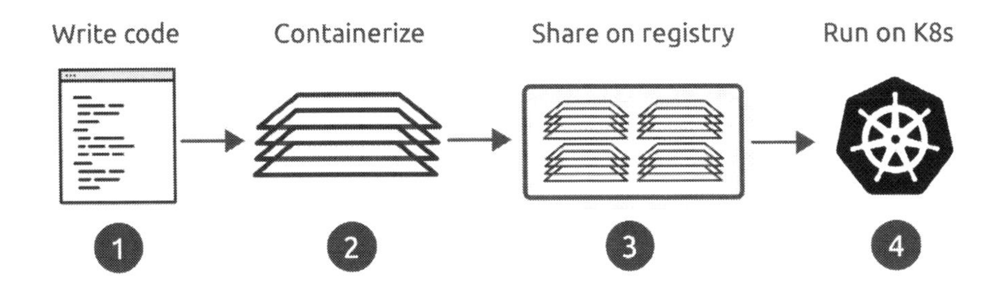

Figure 5.1

The chapter is divided as follows:

- Pre-requisites
- The sample app
- Containerize the app
- Host the image on a registry

Pre-requisites

You'll need all of the following to complete this chapter:

- A copy of the sample app
- Docker
- A Docker account

The sample app is in the book's GitHub repo. If you haven't already done so, run the following command to download it to your computer. It creates a new folder in your current directory and copies the sample app to it.

```
$ git clone https://github.com/nigelpoulton/qsk-book.git
Cloning into 'qsk-book'...
```

Change into the **qsk-book** directory and run an `ls` command to make sure the files are there.

```
$ cd qsk-book

$ ls -l
drwxr-xr-x  7 nigelpoulton  staff  224  App
drwxr-xr-x  8 nigelpoulton  staff  256  Appv1.1
-rw-r--r--  1 nigelpoulton  staff  390  deploy.yml
-rw-r--r--  1 nigelpoulton  staff  225  pod.yml
-rw-r--r--  1 nigelpoulton  staff  929  readme.md
-rw-r--r--  1 nigelpoulton  staff  509  rolling-update.yml
-rw-r--r--  1 nigelpoulton  staff  217  svc.yml
```

Chapter 4 showed you how to get Docker.

You'll need a Docker account if you want to host the containerized app on Docker Hub. *Personal accounts* are free and enable you to follow along. Point your web browser to **hub.docker.com** and complete the sign-up form.

The sample app

The sample app is a simple Node.js web app.

Change into the **App** folder (qsk-book/App) and list the files.

```
$ cd App

$ ls -l
Dockerfile
app.js
bootstrap.css
package.json
views
```

These files make up the application, and it's good to know a bit about each one.

- **Dockerfile:** Contains instructions telling Docker how to containerize the app
- **app.js:** Main application file
- **bootstrap.css:** Stylesheet template that determines how the application's web page looks
- **package.json:** Lists dependencies
- **views:** Folder containing HTML to populate the app's web page

The file of most interest to us in containerizing the app is the *Dockerfile*. It contains instructions Docker uses to build the app into a container image. Ours is simple and looks like this.

```
FROM node:current-slim
LABEL MAINTAINER=nigelpoulton@hotmail.com
COPY . /src
RUN cd /src; npm install
EXPOSE 8080
CMD cd /src && node ./app.js
```

Let's step through it.

The **FROM** instruction tells Docker to pull the **node:current-slim** image from Docker Hub and use it as the base layer for the new image. It contains a minimal Linux OS with *Node* already installed.

LABEL instructions let us add metadata to the image.

The **COPY** instruction tells Docker to copy all the files in the same directory as the Dockerfile into the `/src` folder of the new image. This will copy all the app files and the files listing dependencies.

The **RUN** instruction tells Docker to run an **npm install** command from within the `/src` directory of the new image. This installs the dependencies listed in **package.json**.

The **EXPOSE** instruction adds metadata to the image documenting the application's network port. This matches the port specified in the `app.js` file.

The **CMD** instruction tells Kubernetes what command to run when it creates the container.

In summary, the Dockerfile tells Docker to base the image on the **node:current-slim** image, copy in the app code, install dependencies, and document the network port and the app.

Once you've cloned the repo, you can containerize the app.

Containerize the app

Containerizing an app is the process of packaging the application, and all dependencies, into a *container image*. When the process is complete, the app is said to be *containerized*.

The terms *container image* and *containerized app* mean the same thing.

Use the following `docker build` command to containerize the app. A few quick things to note.

- Run the command from the directory with the Dockerfile
- Substitute **nigelpoulton** with your own Docker account ID
- Include the dot (".") at the end of the command
- The command reads the Dockerfile one line at a time, starting from the top

If you don't have a Docker account, run the command exactly as it is.

```
$ docker build -t nigelpoulton/qsk-book:1.0 .

[+] Building 66.9s (7/7) FINISHED                        0.1s
<Snip>
=> naming to docker.io/nigelpoulton/qsk-book:1.0         0.0s
```

You'll have a new image containing the app and its dependencies. This is the containerized app.

Use the following command to list it. Yours will have a different name, and the output may display other images on your system.

```
$ docker images
REPOSITORY               TAG    IMAGE ID       CREATED          SIZE
nigelpoulton/qsk-book    1.0    c5f4c6f43da5   19 seconds ago   84MB
```

If you're running Docker Desktop, you might see multiple images labelled **registry.k8s.io**. These are system images used to run the built-in Kubernetes cluster.

Now that you've successfully *containerized* the app, the next step is to host it in a registry.

Host the image on a registry

This section is optional, and you'll need a Docker account if you want to follow along. If you don't follow along, you can use a pre-created image in the later chapters.

Container registries are centralised places to store and retrieve images.

Lots of registries exist. Some are on the internet, whereas others are private and can be on your own private network. We'll use Docker Hub as it's the most popular and easiest to use.

Run the following command to push your new image to Docker Hub. Remember to substitute **nigelpoulton** with your own Docker account username. The operation will fail if you use mine, as you don't have permission to push images to my repositories.

```
$ docker push nigelpoulton/qsk-book:1.0

The push refers to repository [docker.io/nigelpoulton/qsk-book]
05a49feb9814: Pushed
66443c37f4d4: Pushed
101dc6329845: Pushed
dc8a57695d7b: Pushed
7466fca84fd0: Pushed
<Snip>
c5f4c6f43da5: Pushed
1.0: digest: sha256:c5f4c6f43da5f0a...aee06961408 size: 1787
```

Go to **hub.docker.com** and make sure the image is present. Remember to browse your own repos.

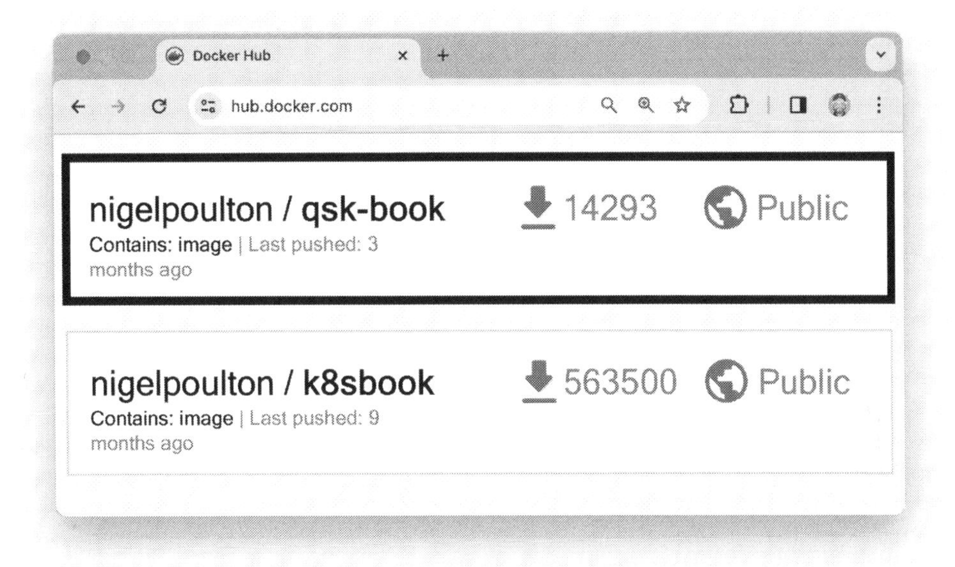

Figure 5.2

At this point, you've containerized the application and pushed it to Docker Hub. This means you're ready to deploy it to Kubernetes.

Chapter summary

In this chapter, you learned that a *containerized app* is just a regular app that's built and packaged as a container image.

You used the **git** CLI to *clone* the book's GitHub repo into a local folder on your computer. You then used Docker to containerize the app and push it to Docker Hub. Along the way, you learned that a *Dockerfile* is a list of instructions telling Docker how to containerize an app.

6: Running an app on Kubernetes

In this chapter, you'll deploy a containerized application to Kubernetes.

If you've been following along, you'll deploy the app you containerized in the previous chapter. If you skipped that, you can use the publicly available image we'll show later.

The chapter is divided as follows:

- Pre-reqs
- Deploy the app
- Test the app

Pre-reqs

You need a working Kubernetes cluster with **kubectl** correctly configured. See Chapter 3.

If you're using Docker Desktop on Windows, you'll need to be in ***Linux containers*** mode. Just right-click the Docker whale in the system tray and choose *Switch to Linux containers*.

Run the following command to make sure kubectl is talking to the right cluster.

```
$ kubectl get nodes
NAME                STATUS   ROLES    AGE   VERSION
k3s-qsk...d5idm     Ready    <none>   5m    v1.28.2+k3s1
k3s-qsk...jhnpn     Ready    <none>   5m    v1.28.2+k3s1
k3s-qsk...um2r2     Ready    <none>   5m    v1.28.2+k3s1
```

The number of nodes will depend on your cluster. If you're using Docker Desktop's built-in Kubernetes cluster, you'll only see one node called **docker-desktop**. Hosted Kubernetes platforms, such as Civo Kubernetes, only list *worker* *nodes.* This is because *control plane nodes* are managed by the cloud platform and hidden from view.

If the command connects to the wrong cluster and you're running Docker Desktop, you can click the Docker whale icon and select the correct cluster, as shown in Figure 6.1.

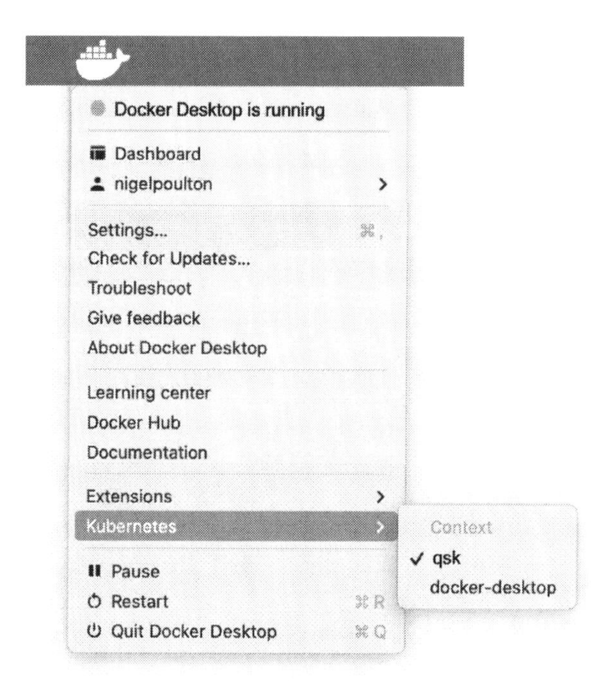

Figure 6.1

If you're not using Docker Desktop and kubectl connects to the wrong cluster, you can change it with the following procedure.

List all contexts defined in your kubeconfig file. A *context* is just a combination of a cluster and an authentication token.

```
$ kubectl config get-contexts
CURRENT   NAME             CLUSTER          AUTHINFO
*         docker-desktop   docker-desktop   docker-desktop
          qsk              qsk              qsk
```

The output lists two contexts. The one with the asterisk (*) is the current context. Your output may be different.

Run the following command to switch to the **qsk** context. You may need to change to a different context.

```
$ kubectl config use-context qsk
Switched to context "qsk".
```

If the command returns the correct nodes and they're showing as *Ready*, you're ready to deploy the app.

Deploy the app to Kubernetes

The first thing to know about deploying containers to Kubernetes is that they have to be wrapped in *Pods*. For now, just think of a *Pod* as a lightweight wrapper with the single goal of allowing containers to be scheduled on Kubernetes.

Figure 6.2 shows a Pod called *first-pod* wrapping a single container called *web*. The Pod is only adding metadata to assist with scheduling.

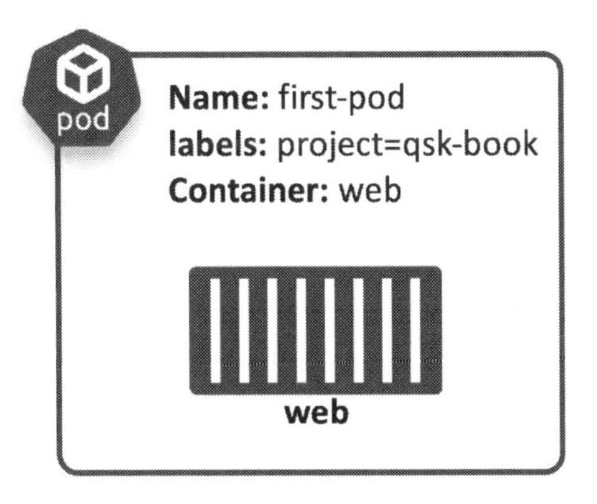

Figure 6.2

Your first Kubernetes Pod

The Pod you'll deploy is defined in a YAML file called **pod.yml** in the root of the book's GitHub repo.

The file can have any name, but the contents follow strict YAML rules. If you don't already know, YAML is a popular language for configuration files and is very strict about indentation.

```
1  kind: Pod
2  apiVersion: v1
3  metadata:
4    name: first-pod
5    labels:
6      project: qsk-book
7  spec:
8    containers:
9      - name: web
10       image: nigelpoulton/qsk-book:1.0
11       ports:
12         - containerPort: 8080
```

This Pod wraps a single container. Lines 1-7 are the Pod, and lines 8-12 define the container it wraps.

Let's have a closer look.

The **kind** and **apiVersion** fields tell Kubernetes the type and version of the object you're deploying. In this case, we're telling Kubernetes to create a new Pod based on version 1 of the Pod specification.

The **metadata** block gives the Pod a name and a label. We'll use the name later to help us identify the Pod when it's running. We'll use the label to connect it to a load balancer.

The **spec** section defines the containers the Pod will run. This Pod runs a single container, called *web,* based on the image created in the previous chapter. You can change this to your own image if you followed along in the previous chapter. If you didn't follow along, leave it as it is.

Figure 6.3 shows the Pod wrapping the container. Remember, Kubernetes will only run containers wrapped in Pods.

Figure 6.3

Deploy the app (Pod)

The recommended way to deploy a new Pod is with the **kubectl apply** command. This sends the Pod's YAML file to Kubernetes, and the control plane takes care of deploying it.

Run the following command to list any existing Pods on your cluster. If you're working with a new cluster, you won't have any.

```
$ kubectl get pods
No resources found in default namespace.
```

Run the following commands from the **qsk-book** folder where the **pod.yml** file is located. If you're currently in the **App** directory (check with pwd), you'll need to back up one directory level with the "`cd ..`" command.

Deploy the Pod with the following command.

```
$ kubectl apply -f pod.yml
pod/first-pod created
```

The command sent the **pod.yml** file to the API server where the request was authenticated and authorized using the credentials from your *kubeconfig* file. After that, Kubernetes persisted the Pod definition to the *cluster store* and the *scheduler* allocated it to a worker node.

Check to see if it's running. It may take a few seconds for Kubernetes to pull the image and start it.

```
$ kubectl get pods
NAME        READY   STATUS    RESTARTS   AGE
first-pod   1/1     Running   0          28s
```

Congratulations, the containerized app is running inside a Pod on your Kubernetes cluster!

Inspect the app

`kubectl` provides the **get** and **describe** commands to query the configuration and state of objects. You've already seen that `kubectl get` provides very basic info. The following example shows a `kubectl describe`, which returns a lot more detail. In fact, the output has been trimmed so it only shows the most relevant parts. Take a minute to look through the output.

```
$ kubectl describe pod first-pod

Name:           first-pod
Namespace:      default
Node:           k3s-qsk-885e/192.168.1.4
Labels:         project=qsk-book
Status:         Running
IP:             10.42.2.3
Containers:
  web:
    Container ID:   containerd://a1ec2a8b7180e1...
```

```
    Image:          nigelpoulton/qsk-book:1.0
    Port:           8080/TCP
    State:          Running
    <Snip>
Conditions:
  Type              Status
  Initialized       True
  Ready             True
  ContainersReady   True
  PodScheduled      True
Events:
  Type    Reason     Age   From     Message
  ----    ------     ----  ----     -------
  <Snip>
  Normal  Started    83s   kubelet  Started container web
```

The Pod is up, and it has an IP address. However, this is an internal IP address that's not accessible from outside the cluster. For us to access the app, we need to put a *Service* in front of the Pod and use the Service to access it.

Connect to the app

We always connect to Pods through a Service.

A Service is a Kubernetes object designed to provide stable networking for Pods. As shown in Figure 6.4, they have a front-end and a back-end. The front-end provides a name, IP, and a port that clients send requests to. The back-end forwards these requests to Pods with matching labels.

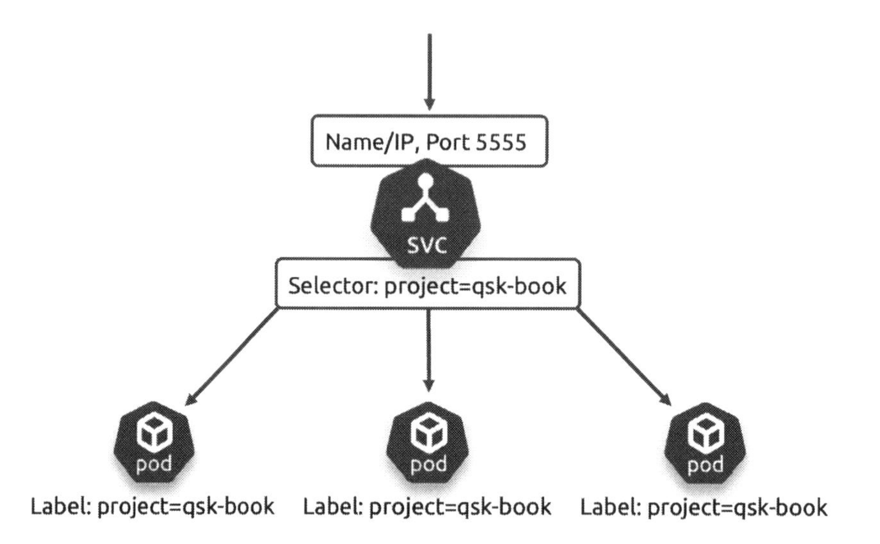

Figure 6.4

We often use the term *object* when referring to things running on Kubernetes. You've already deployed a Pod *object,* and you're about to deploy a Service *object*. We also capitalise the first letter of Kubernetes objects so that it's clear we're referring to the Kubernetes objects and not something else. For example, *Pod, Service, Deployment,* and *Ingress* are all names of Kubernetes objects that have other meanings outside of Kubernetes. By capitalising the first letter, you can be sure we're referring to the Kubernetes object.

Your first Kubernetes Service

We'll deploy the Service defined in the **svc.yml** file in the root folder of the book's GitHub repo.

Here's what the file looks like.

```
1  kind: Service
2  apiVersion: v1
3  metadata:
4    name: svc-lb
5  spec:
6    type: LoadBalancer
7    ports:
8    - port: 5555
9      targetPort: 8080
10   selector:
11     project: qsk-book
```

Let's step through it.

The first two lines are similar to the Pod YAML. They tell Kubernetes to deploy a Service object using the v1 specification.

The **metadata** section names the Service **svc-lb**. Other Pods on the cluster can connect to this name and access the Pods behind it.

The **spec** section is where the magic happens. This one defines a *LoadBalancer* Service that accepts traffic on port 5555 and forwards it on port 8080 to any Pods with the **project=qsk-book** label. If your cluster is on the Civo Cloud, it will provision one of Civo's internet-facing load balancers. If your cluster is on Docker Desktop, it will be accessible on **localhost**.

Figure 6.5 shows a Kubernetes cluster on the Civo Cloud. The app in the Pod is listening on port 8080 and fronted by the **svc-lb** Service. As this is a *LoadBalancer* Service, it creates an internet-facing load balancer on the Civo cloud that listens on the internet on port 5555. Clients hitting the load balancer on port 5555 will be routed to the Pod.

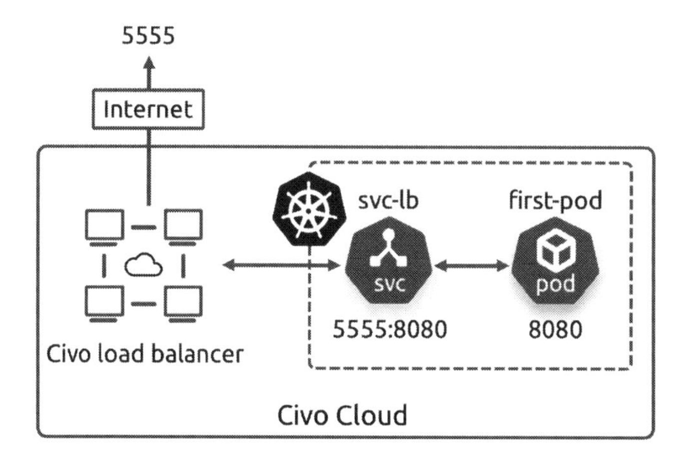

Figure 6.5

A quick word on labels

Labels are the main way that Kubernetes connects objects.

If you look closely at the `pod.yml` and `svc.yml` files, you'll see they both reference the **project: qsk-book** label.

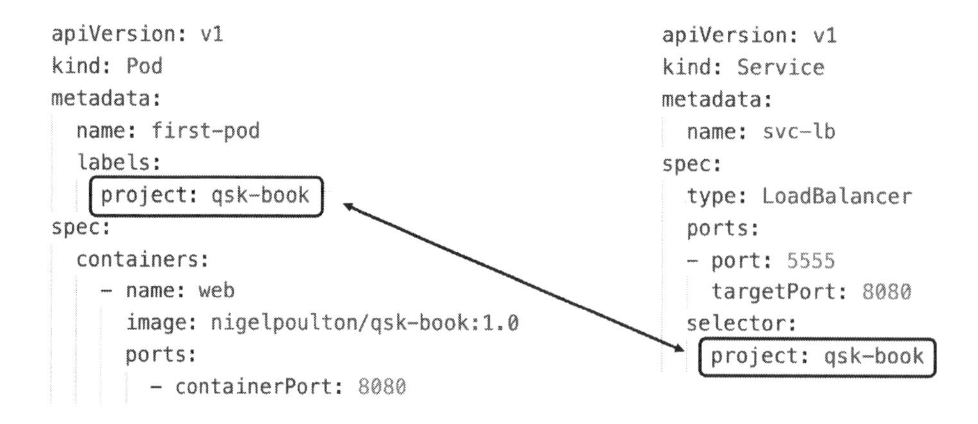

Figure 6.6

The Service accepts traffic on port 5555 and forwards it, on port 8080, to *all Pods with the **project=qsk-book** label.* It also maintains an up-to-date list of Pods with the label.

Currently, only one Pod has the label. However, if you add more, Kubernetes will forward traffic to them all. You'll see this in the next chapter.

Deploy the Service

As with Pods, you deploy Services with `kubectl apply`.

Run the following command to deploy the Service. Be sure to run it from the **qsk-book** folder where the **svc.yml** file is located.

```
$ kubectl apply -f svc.yml
service/svc-lb created
```

Verify the Service is up and running. You can also run a `kubectl describe svc svc-lb` command to get more detailed info.

```
$ kubectl get svc
NAME    TYPE           CLUSTER-IP    EXTERNAL-IP   PORT(S)
svc-lb  LoadBalancer   10.43.63.98   74.220.22.94  5555:31384/TCP
```

Your output may show `<pending>` in the `EXTERNAL-IP` column while your cloud provisions an internet-facing load balancer. This can take a few minutes on some cloud platforms. It will show **localhost** if you're using the built-in Docker Desktop cluster.

Let's look closer at the output.

The Service is called *svc-lb*, and the type is correctly set as *LoadBalancer*.

The **CLUSTER-IP** is the Service's internal IP address on the Pod network. This is what Pods on the same cluster will use to access it.

The **EXTERNAL-IP** is the address we'll use to connect to the app. If you're following along on a cloud, this will be a public IP you can access from the internet. If you're using Docker Desktop, it'll be *localhost* and only accessible from your local machine.

The **PORT(S)** column lists the ports the Service is accessible on. We'll use the first value, 5555.

Point a browser to the address in the **EXTERNAL-IP** column on port 5555 to see the app.

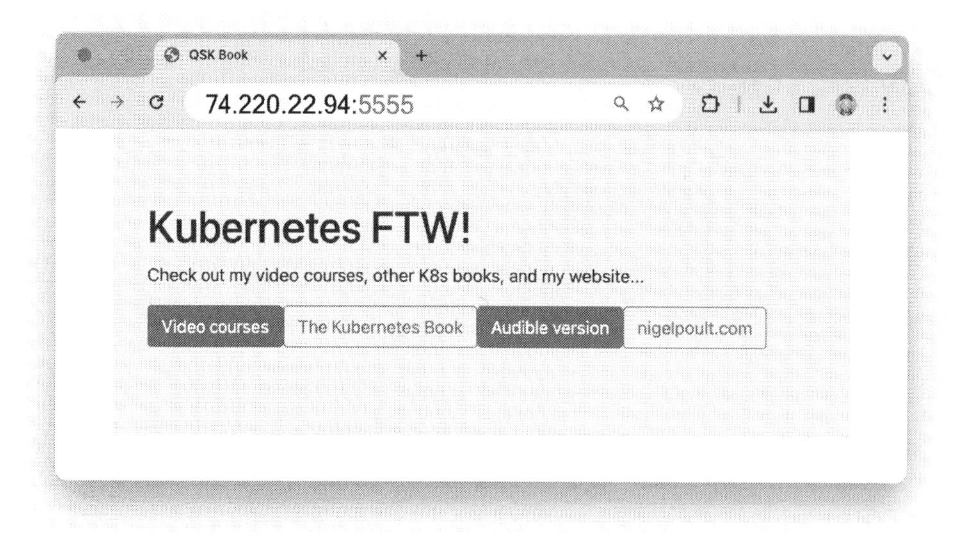

Figure 6.7

Congratulations, you've containerized an app, deployed it to Kubernetes in a Pod, used a Service to provision an internet-facing load balancer, and connected to the app.

Clean-up

Let's tidy things up so you've got a clean system for the start of the next chapter.

Run the following commands to delete the Service and Pod. It may take a few seconds for the Pod to terminate while it waits for the app to gracefully shutdown.

```
$ kubectl delete svc svc-lb
service "cloud-lb" deleted

$ kubectl delete pod first-pod
pod "first-pod" deleted
```

Chapter summary

In this chapter, you learned that containerized apps have to be wrapped in Pods if they want to run on Kubernetes. Fortunately, Pods are lightweight constructs and add no overhead.

You saw a simple Pod defined in a YAML file and learned how to deploy it to Kubernetes with the `kubectl apply` command. You also learned how to inspect objects with `kubectl get` and `kubectl describe`.

Finally, you learned that Kubernetes Services allow you to connect to apps running in Pods.

So far, you've built, deployed, and connected to a containerized app. However, you haven't seen self-healing, scaling, or any other cloud-native features. You'll see all of these in the upcoming chapters.

7: Self-healing

In this chapter, you'll learn about the Kubernetes *Deployment* object and use it to make the sample app *resilient.*

The chapter is organised as follows:

- Intro to Kubernetes Deployments
- Self-heal from a Pod failure
- Self-heal from a worker node failure

Intro to Kubernetes Deployments

Kubernetes has lots of *objects* that add features and capabilities. In the previous chapter, you used a *Service* object to provide network connectivity. In this chapter, you'll use a *Deployment* object to add self-healing. In fact, in the next few chapters, you'll use the same Deployment object to perform scaling and rolling updates.

As with Pods and Services, *Deployments* are defined in YAML manifest files.

Figure 7.1 shows the Deployment manifest we'll use. It's marked up to show the container nested in the Pod and the Pod nested in the Deployment.

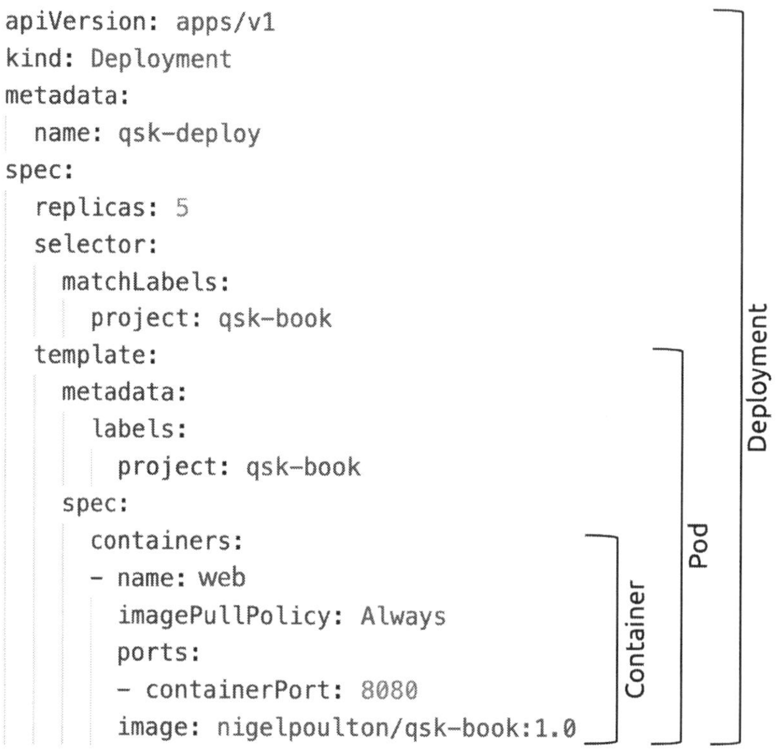
apiVersion: apps/v1
kind: Deployment
metadata:
  name: qsk-deploy
spec:
  replicas: 5
  selector:
    matchLabels:
      project: qsk-book
  template:
    metadata:
      labels:
        project: qsk-book
    spec:
      containers:
      - name: web
        imagePullPolicy: Always
        ports:
        - containerPort: 8080
        image: nigelpoulton/qsk-book:1.0
Container
Pod
Deployment

Figure 7.1

This nesting, or wrapping, is important in understanding how each level adds something else:

- The application is at the center
- The container adds the OS and app dependencies
- The Pod adds metadata so it can be scheduled on Kubernetes
- The Deployment adds cloud-native features such as self-healing, scaling, and rollouts

How Deployments work

There are two main parts to Deployments.

1. The object
2. The controller

At the highest level, the *object* holds the definition, whereas the *controller* implements it and makes sure it runs properly.

Consider a quick example.

You have a Deployment YAML that defines a Pod and requests five replicas. You use kubectl to post it to the API server, and five Pods get scheduled to the cluster.

At this point, *desired state* and *observed state* are in sync — you asked for five Pods, and you've got five Pods. But let's say a node fails, and you drop from five Pods to four. *Observed state* no longer matches *desired state.*

Without Kubernetes, this would be a problem, and somebody would need to fix it. However, the Kubernetes Deployment *controller* is constantly watching the cluster and will notice the change. It knows you *desire* five Pods, but it can only *observe* four. So, it'll start a new Pod to bring *observed state* back into sync with *desired state.*

The technical term for this is *reconciliation,* but we often call it *self-healing.*

Let's test it.

Self-heal from a Pod failure

In this section, you'll use a Kubernetes Deployment to deploy five replicas of a Pod. After that, you'll manually delete a Pod and see Kubernetes self-heal.

You'll use the **deploy.yml** file in the root of the book's GitHub repo. It defines five replicas of the app you containerized in previous chapters.

```
kind: Deployment                  <<== Type of object being defined
apiVersion: apps/v1               <<== Version of object to deploy
metadata:
  name: qsk-deploy
spec:
  replicas: 5                     <<== How many Pod replicas
  selector:
    matchLabels:                  <<== Tells the Deployment controller
      project: qsk-book           <<== to manage Pods with this label
  template:
    metadata:
      labels:
        project: qsk-book         <<== Give all replicas this label
    spec:
      containers:
      - name: qsk-pod
        imagePullPolicy: Always  <<== Never use images from local machine
        ports:
        - containerPort: 8080          <<== Network port
        image: nigelpoulton/qsk-book:1.0  <<== Container image to use
```

We use the terms *Pod, instance,* and *replica* to mean the same thing — a Pod running a containerized app.

Check your cluster for any existing Pods and Deployments.

```
$ kubectl get pods
No resources found in default namespace.

$ kubectl get deployments
No resources found in default namespace.
```

Now use `kubectl` to deploy the Deployment to your cluster. Run the command from the same folder as the **deploy.yml** file.

```
$ kubectl apply -f deploy.yml
deployment.apps/qsk-deploy created
```

Check the status of the Deployment and the Pods it's managing.

```
$ kubectl get deployments
NAME         READY   UP-TO-DATE   AVAILABLE   AGE
qsk-deploy   5/5     5            5           21s

$ kubectl get pods
NAME                     READY   STATUS    RESTARTS   AGE
qsk-deploy-84...5txzv    1/1     Running   0          31s
qsk-deploy-84...mbscc    1/1     Running   0          31s
qsk-deploy-84...mr4d8    1/1     Running   0          31s
qsk-deploy-84...nwr6z    1/1     Running   0          31s
qsk-deploy-84...whsnt    1/1     Running   0          31s
```

You requested five replicas, and you've got five replicas. This means observed state matches desired state, and there's no more work for the Deployment controller to do. However, it keeps running in the background watching the state of the Deployment.

Pod failure

As with all software, Pods can crash and fail. However, if they're managed by a Deployment controller, Kubernetes will *attempt* to recover from failures by starting new Pods to replace failed ones.

Run a `kubectl delete pod` command to manually delete one of the Pods. You'll need to use a Pod name from your environment, and it may take a few seconds for the Pod to delete.

```
$ kubectl delete pod qsk-deploy-845b58bd85-5txzv
pod "qsk-deploy-845b58bd85-5txzv" deleted
```

As soon as the Pod is deleted, *observed state* will drop to four and no longer match the *desired state* of five. The Deployment controller will notice this and automatically start a new one, taking the observed number of Pods back to five.

List the Pods again to see if a new one has been started.

```
$ kubectl get pods
NAME                          READY   STATUS    RESTARTS   AGE
qsk-deploy-845b58bd85-2dx4s   1/1     Running   0          35s
qsk-deploy-845b58bd85-mbscc   1/1     Running   0          2m51s
qsk-deploy-845b58bd85-mr4d8   1/1     Running   0          2m51s
qsk-deploy-845b58bd85-nwr6z   1/1     Running   0          2m51s
qsk-deploy-845b58bd85-whsnt   1/1     Running   0          2m51s
```

Notice how the first Pod in the list has only been running for 35 seconds. This is the new one Kubernetes automatically created to *reconcile desired state.*

Congratulations. You just simulated a Pod failure, and Kubernetes self-healed without needing help.

Let's see how Kubernetes deals with the *node* failures.

Self-heal from a worker node failure

When a worker node fails, all Pods running on the node are lost. However, if a Deployment controller manages the Pods, Kubernetes will start replacements on surviving nodes.

If your cluster is on a cloud that implements *node pools,* the failed **node** will also be replaced. However, this is a feature of your cloud's hosted Kubernetes service and not a feature of Deployments — Deployments do not heal nodes, they only heal Pods.

You can only follow the steps in this section if you have a multi-node cluster and can delete worker nodes. If you built a multi-node cluster on the Civo Cloud, as explained in Chapter 3, you can follow along. If you're using the single-node Docker Desktop cluster, you can't follow along.

The following command lists all Pods and the worker nodes they're running on. The output is trimmed to fit the book.

```
$ kubectl get pods -o wide
NAME           READY   STATUS    <Snip>    NODE
qsk...2dx4s    1/1     Running   ...       k3s...6s
qsk...mbscc    1/1     Running   ...       k3s...2n
qsk...mr4d8    1/1     Running   ...       k3s...2n
qsk...nwr6z    1/1     Running   ...       k3s...rq
qsk...whsnt    1/1     Running   ...       k3s...6s
```

See how the five Pods are spread across three worker nodes.

The next step will delete the **k3s...2n** worker node and kill the two Pods running on it.

The following process shows how to delete a worker node on Civo Cloud Kubernetes. Deleting it this way simulates sudden node failure.

1. Open your Kubernetes cluster in the Civo dashboard
2. Scroll down to **Node Pools**
3. Click the **X** on the node you wish to delete
4. Click **Recycle node**

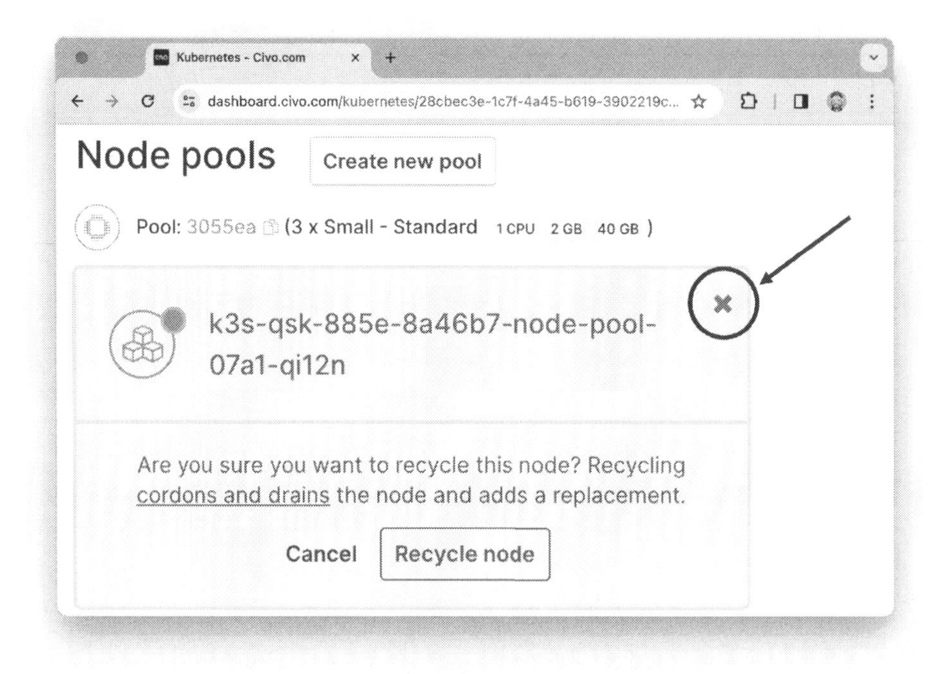

Figure 7.2

The following command verifies the node has been deleted. If you wait too long to run it, Civo will have already replaced the deleted node. It can take a minute or two for the missing node to disappear from the command output.

```
$ kubectl get nodes
NAME                STATUS    ROLES     AGE    VERSION
k3s...6s            Ready     <none>    23h    v1.28.2+k3s1
k3s...rq            Ready     <none>    23h    v1.28.2+k3s1
```

The node is deleted, and the output only returns the two remaining healthy nodes.

As soon as the Deployment controller notices the missing Pods, it creates replacements. Run the following command to verify this. It may take a few seconds for the replacement Pods to reach the Running state.

```
$ kubectl get pods -o wide
NAME          READY   STATUS              <Snip>   NODE
qsk...2dx4s   1/1     Running             ...      k3s...6s
qsk...nwr6z   1/1     Running             ...      k3s...rq
qsk...whsnt   1/1     Running             ...      k3s...6s
qsk...nwmwx   1/1     ContainerCreating   ...      k3s...rq
qsk...rl2p2   1/1     ContainerCreating   ...      k3s...6s
```

The output shows Kubernetes creating two new Pods to replace the two that were lost when the **k3s...2n** node was deleted. Both have been scheduled to surviving worker nodes.

You can verify the state of the Deployment with the following command.

```
$ kubectl get deployments
NAME         READY   UP-TO-DATE   AVAILABLE   AGE
qsk-deploy   5/5     5            5           38m
```

Congratulations. You've simulated a node failure, and Kubernetes automatically replaced the Pods that were lost.

After a few more minutes, Civo Cloud will replace the deleted node and return the cluster to three worker nodes. This is a feature of Civo Cloud and not the Kubernetes Deployment controller. It works because Civo Cloud's implementation of *node pools* also has the notion of *desired state*. When you created the cluster, you requested three worker nodes. When one was deleted, the Civo Cloud noticed the change and added a new one to bring observed state back into sync with desired state.

Although the cluster is back to three worker nodes, Kubernetes won't attempt to re-balance existing Pods across all available nodes. This means you'll have a three-node cluster with all Pods running on just two nodes.

Chapter summary

In this chapter, you learned that Kubernetes has a *Deployment object* that works with the *Deployment controller* to implement self-healing of Pods. The Deployment controller runs on the control plane, ensuring *observed state* matches *desired state*.

You also saw how Deployments wrap Pods, which in turn wrap containers, which in turn wrap applications.

You used **kubectl** to deploy an app via a Deployment object and tested self-healing. You manually deleted a Pod and a worker node and watched Kubernetes self-heal from both failures.

Civo Cloud Kubernetes also replaced the deleted/broken worker node.

8: Scaling the app

In this chapter, you'll scale the app up and down. The methods you'll learn are *manual* and require a human to implement them. In the real world, you'll use a *Horizontal Pod Autoscaler (HPA)* to make scaling automatic. These are beyond the scope of a quick start book, but the things you'll learn here will be extremely valuable.

However, Kubernetes also has objects that enable automatic scaling, but they're beyond the scope of a quick start book.

We've split the chapter as follows.

- Pre-requisites
- Scale an application up
- Scale an application down
- The role of labels
- Declarative vs imperative

Pre-requisites

If you've been following along, you'll already have a single Deployment managing five replicas of the app you containerized in Chapter 5. If you already have this Deployment running, you can skip to the **Scale an application up** section.

If you haven't been following along, run the following command to deploy five replicas of a containerized app. You'll need a working cluster, and be sure to run the command from the root of the book's GitHub repo where the **deploy.yml** file is located.

```
$ kubectl apply -f deploy.yml
deployment.apps/qsk-deploy created
```

Run the following command to make sure the Deployment is running and managing five replicas.

```
$ kubectl get deployments
NAME         READY   UP-TO-DATE   AVAILABLE   AGE
qsk-deploy   5/5     5            5           52s
```

You can move to the next section as soon as all five replicas are up and running.

Scale an application up

In this section, you'll edit the Deployment YAML file, increase the replica count to ten, and re-send the file to Kubernetes. This will kick off the *reconciliation* process and increase the number of replicas on your cluster to ten.

Before doing this, it's important to know that the unit of scaling in Kubernetes is the Pod. This means Kubernetes adds Pods to scale up and deletes Pods to scale down.

Check the current number of replicas.

```
$ kubectl get deployment qsk-deploy
NAME         READY   UP-TO-DATE   AVAILABLE   AGE
qsk-deploy   5/5     5            5           2m5s
```

Use your favorite editor to edit the `deploy.yml` file and set the `spec.replicas` field to 10 and **save your changes**.

```
apiVersion: apps/v1
kind: Deployment
metadata:
  name: qsk-deploy
spec:
  replicas: 5            <<== Change this to 10
  selector:
    matchLabels:
      project: qsk-book
<Snip>
```

Be sure you've saved your changes.

Three important things will happen when you re-send the file to Kubernetes:

1. The *desired state* will change from five replicas to ten
2. The Deployment controller will observe the five replicas and realise it doesn't match the desired state of ten
3. The Deployment controller will schedule five new replicas to increase the total number to ten

This is the exact same *reconciliation* process you saw when Kubernetes self-healed from Pod failures. *Observed state* didn't match *desired state,* so Kubernetes fixed it.

Run the following command to send the updated file to the API server.

```
$ kubectl apply -f deploy.yml
deployment.apps/qsk-deploy configured
```

Run a couple of commands to check the status of the Deployment and that it's now managing **ten** Pods.

```
$ kubectl get deployment qsk-deploy
NAME         READY   UP-TO-DATE   AVAILABLE   AGE
qsk-deploy   10/10   10           10          3m55s

$ kubectl get pods
NAME                          READY    STATUS    RESTARTS    AGE
qsk-deploy-845b58bd85-2dx4s   1/1      Running   0           3m55s
qsk-deploy-845b58bd85-5pls9   1/1      Running   0           21s
qsk-deploy-845b58bd85-5rfsj   1/1      Running   0           21s
qsk-deploy-845b58bd85-66hzd   1/1      Running   0           21s
qsk-deploy-845b58bd85-ffn4t   1/1      Running   0           21s
qsk-deploy-845b58bd85-hjc5x   1/1      Running   0           21s
qsk-deploy-845b58bd85-mbscc   1/1      Running   0           3m55s
qsk-deploy-845b58bd85-mr4d8   1/1      Running   0           3m55s
qsk-deploy-845b58bd85-nwr6z   1/1      Running   0           3m55s
qsk-deploy-845b58bd85-whsnt   1/1      Running   0           3m55s
```

The new Pods might take a few seconds to start, but you can easily identify them based on their age.

If you followed the examples in the previous chapter and deleted a node, most of the new Pods will be scheduled on the new node. This is Kubernetes trying to balance the Pods across all the worker nodes in the cluster.

Congratulations. You've manually scaled the application from five replicas to ten using the *declarative method.* This is jargon for *declaring* a new desired state in the YAML file and using the file to update the cluster.

Scale an application down

In this section, you'll use the **kubectl scale** command to scale the number of Pods back down to five. This is called the *imperative method* and is not as recommended as the *declarative method.*

Run the following command.

```
$ kubectl scale --replicas 5 deployment/qsk-deploy
deployment.apps/qsk-deploy scaled
```

Check the number of Pods. As always, deleted Pods can take a few seconds to fully terminate.

```
$ kubectl get pods
NAME                            READY   STATUS    RESTARTS   AGE
qsk-deploy-845b58bd85-2dx4s     1/1     Running   0          6m
qsk-deploy-845b58bd85-mbscc     1/1     Running   0          6m
qsk-deploy-845b58bd85-mr4d8     1/1     Running   0          6m
qsk-deploy-845b58bd85-nwr6z     1/1     Running   0          6m
qsk-deploy-845b58bd85-whsnt     1/1     Running   0          6m
```

Congratulations. You've manually scaled the application back down to five replicas.

The role of labels

Deployments use labels to ensure they only ever manage the Pods they created.

Every time a Deployment creates a Pod, it gives the new Pod a label. It then uses this label to know which Pods it created and can manage.

For example, Figure 8.1 shows a single Deployment and five Pods. However, only four of the Pods were created by the Deployment and have its label. The one on the far right has a different label and isn't managed by the Deployment.

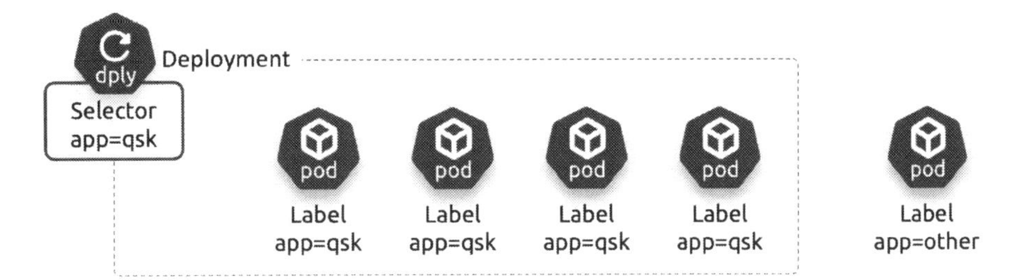

Figure8.1

Consider a couple of quick examples.

If you scale the Deployment down from four Pods to two, it will only delete Pods with the **app=qsk** label.

If the Deployment is scaled up, it will add more Pods with the same label. This means existing *Services* sending traffic to the Deployment's Pods will automatically start sending traffic to the new Pods as they'll have the same label. This is shown in Figure 8.2

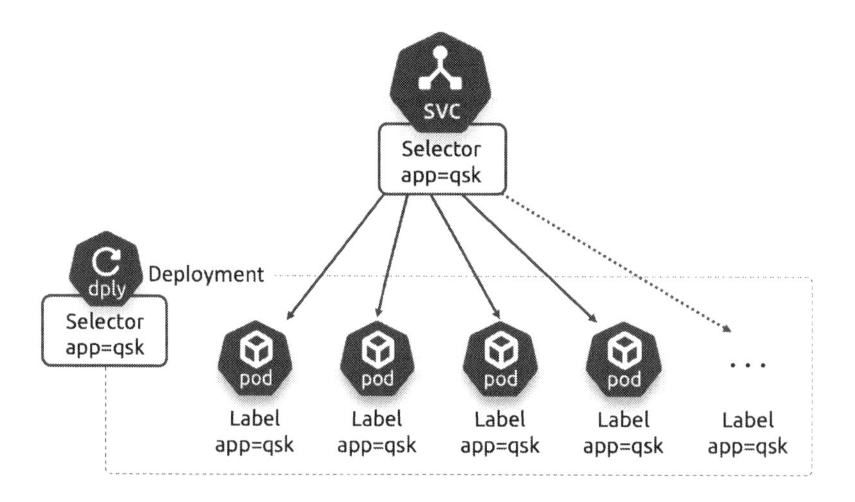

Figure8.2

Declarative vs imperative

You've seen two ways to perform updates:

- Declaratively
- Imperatively

The *declarative* method is the preferred method and requires all updates to be performed through YAML configuration files.

The *imperative* method uses kubectl commands to perform updates and isn't recommended for live production environments.

Consider the following example.

You have a Deployment YAML file defining four replicas of a Pod that you deploy to your cluster. Everything is fine until demand increases and the app starts responding slowly. Somebody comes along and uses the ***kubectl scale*** *command to increase the number of replicas from four to eight. This fixes the slow response times, but the state of the cluster and the YAML file are no longer in sync — the cluster is running eight replicas, but the Deployment YAML file only defines four.*

Sometime later, you need to push a new version of the app. To accomplish this, you open the YAML file, update the image version it references, save your changes, and re-post the file to the cluster. This successfully updates the version of the image, but it also decreases the number of replicas to four!

If demand is still high, the app will start responding slowly again, and you may think the problem is with the new version and not realise you accidentally reduced the number of replicas.

Reasons like this are why it's considered a good practice to manage everything declaratively.

Edit your **deploy.yml** file, set the number of replicas back to five and save your changes. It now matches what is deployed to your cluster.

Chapter summary

In this chapter, you learned how to manually scale a Deployment by editing its YAML file and re-sending it to Kubernetes. This is called the *declarative* method. You also saw that it's possible to perform scaling operations using the `kubectl scale` command. This is the *imperative* method and not recommended.

9: Performing a rolling update

In this chapter, you'll perform a *zero-downtime rolling update* on the app we've been working with in the previous chapters. If you're unsure what a zero-downtime rolling update is, great, you're about to find out.

We'll divide this chapter as follows.

- Pre-requisites
- Performing a rollout

You can follow all the examples in this chapter using either the Docker Desktop built-in Kubernetes cluster or a Civo Cloud Kubernetes cluster we showed you how to get in Chapter 3. You can also use other setups.

Pre-requisites

If you've followed along in previous chapters, you'll already have the **qsk-deploy** Deployment running on your cluster managing five replicas. If you do, skip to the *Performing a rollout* section.

If you haven't followed along, complete these steps to get ready.

1. Get a Kubernetes cluster and configure `kubectl` (see Chapter 3)
2. Clone the book's GitHub repo
3. Deploy the sample app and Service

If you haven't already done so, clone the book's GitHub repo and change into the **qsk-book** folder.

```
$ git clone https://github.com/nigelpoulton/qsk-book.git
Cloning into 'qsk-book'...

$ cd qsk-book
```

Run the following command to deploy the app and the load balancer Service. Be sure to run it from the **qsk-book** folder.

```
$ kubectl apply -f deploy.yml -f svc.yml
deployment.apps/qsk-deploy created
service/svc-lb created
```

Run a `kubectl get deployments` and `kubectl get svc` command to make sure the application and Service are both running.

```
$ kubectl get deployments
NAME         READY   UP-TO-DATE   AVAILABLE   AGE
qsk-deploy   5/5     5            5           18s

$ kubectl get svc
NAME     TYPE           CLUSTER-IP      EXTERNAL-IP    PORT(S)
svc-lb   LoadBalancer   10.43.156.109   74.220.16.41   5555:31773/TCP
```

It may take a minute for all five Pods to enter the *ready* state and the load balancer Service to get a public IP (*"localhost"* on Docker Desktop). Once you have these, proceed to the next section.

Deploy the rollout

In this section, you'll perform a rolling update so that all five replicas run a new version of the app. You'll force Kubernetes to update one replica at a time, with a short pause between each.

On the jargon front, we use the terms *rollout, update,* and *rolling update* to mean the same thing.

Figure 9.1 shows the high-level process of updating a running app to a new version. Steps 1-3 have already been completed, and we'll focus on step 4.

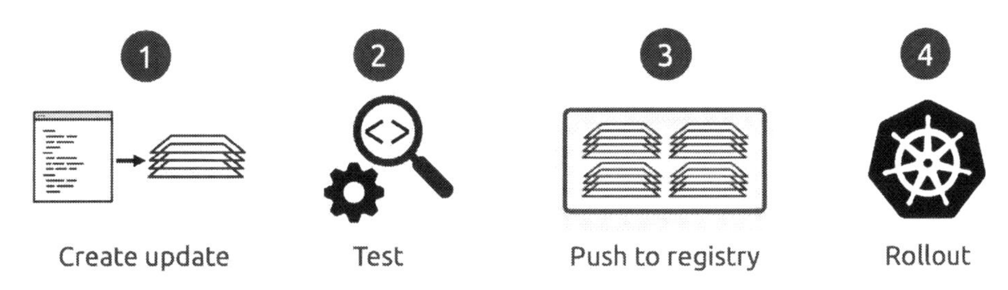

Figure 9.1

You'll complete the following:

1. Edit the **deploy.yml** file to configure the update settings and specify the new version of the image
2. Re-send the updated YAML file to Kubernetes
3. Observe the update process
4. Test the new version

Edit the Deployment YAML file

Open the **deploy.yml** file and make all the changes shown in the following snippet.

- Lines 10-15 tell Kubernetes how to perform the update (we'll explain these shortly)
- Line 26 specifies the updated version of the image that is already built and pushed to Docker Hub

```
1 apiVersion: apps/v1
2 kind: Deployment
3 metadata:
4   name: qsk-deploy
5 spec:
6   replicas: 5
7   selector:
8     matchLabels:
9       project: qsk-book
10  minReadySeconds: 20        <<== Add this line
11  strategy:                  <<== Add this line
12    type: RollingUpdate      <<== Add this line
13    rollingUpdate:           <<== Add this line
14      maxSurge: 1            <<== Add this line
15      maxUnavailable: 0      <<== Add this line
16  template:
17    metadata:
18      labels:
19        project: qsk-book
20    spec:
21      containers:
22      - name: hello-pod
23        imagePullPolicy: Always
24        ports:
25        - containerPort: 8080
26        image: nigelpoulton/qsk-book:1.1    <<== Set to 1.1
```

Before going any further. YAML is strict about proper indentation. This means you need to be extra careful that you've indented each new line the correct number of ***spaces***. Also, you cannot mix-and-match tabs and spaces in the same file.

If you have issues editing the file, you can use a pre-completed version called **rolling-update.yml** instead.

Be sure to save your changes.

Understand the update settings

You added the following six lines that tell Kubernetes how to perform the update.

```
10  minReadySeconds: 20
11  strategy:
12    type: RollingUpdate
13    rollingUpdate:
14      maxSurge: 1
15      maxUnavailable: 0
```

minReadySeconds, on line 10, tells Kubernetes to wait for 20 seconds after updating each replica. This means Kubernetes will update the first replica, wait 20 seconds, update the second replica, wait 20 seconds, update the third, etc.

Inserting waits like this gives you a chance to run tests and make sure the new replicas are working as expected. In the real world, you'll wait longer than 20 seconds between replica updates.

Also, Kubernetes doesn't actually *update* replicas. It **deletes** existing replicas and replaces them with brand-new ones running the new version.

Lines 11 and 12 force Kubernetes to use the *RollingUpdate* strategy when performing the update. This defaults to updating one replica at a time and is different to the *Recreate* strategy that deletes and replaces all Pods in one go.

Lines 14 and 15 force Kubernetes to update one Pod at a time as follows…

maxSurge=1 gives Kubernetes permission to add one extra Pod during the rollout process. In our case, desired state is five Pods, so this setting allows the rollout to temporarily *surge* to six Pods. **maxUnavailable=0** on line 15 prevents the update from going below five Pods. When combined, lines 14 and 15 force Kubernetes to add a sixth replica with the new version and then delete an existing replica running the old version. This process repeats until all five Pods are running the desired version.

Perform the rolling update

Make sure you've saved the changes, then use `kubectl apply` to send the updated configuration file to Kubernetes.

```
$ kubectl apply -f deploy.yml
deployment.apps/qsk-deploy configured
```

Kubernetes will record a new desired state, of five Pods running version 1.1 of the image, in the cluster store. The Deployment controller will observe the cluster, see that it has five Pods on a different version, and start replacing them, one at-a-time, with a 20-second wait between each.

Monitor and check the rolling update

You can monitor the progress of the job with the following command. The output has been trimmed to fit the page.

If your output looks different, it may be because you waited too long to run the command, and the operation has already completed.

```
$ kubectl rollout status deployment qsk-deploy
Waiting for rollout to finish: 1 out of 5 have been updated...
Waiting for rollout to finish: 1 out of 5 have been updated...
Waiting for rollout to finish: 2 out of 5 have been updated...
Waiting for rollout to finish: 2 out of 5 have been updated...
Waiting for rollout to finish: 3 out of 5 have been updated...
Waiting for rollout to finish: 3 out of 5 have been updated...
Waiting for rollout to finish: 4 out of 5 have been updated...
Waiting for rollout to finish: 4 out of 5 have been updated...
Waiting for rollout to finish: 2 old replicas are pending termination...
Waiting for rollout to finish: 1 old replicas are pending termination...
deployment "qsk-deploy" successfully rolled out
```

You may also be able to point your web browser at the app and keep refreshing the page. Some of your requests might return the original version of the app, whereas others might return the new version. Once all five replicas are up to date, all requests will return the latest version.

Figure 9.2

Congratulations. You've performed a successful rolling update.

Clean-up

The following commands delete the Deployment and Service from your cluster.

```
$ kubectl delete deployment qsk-deploy
deployment.apps "qsk-deploy" deleted

$ kubectl delete svc cloud-lb
service "cloud-lb" deleted
```

If your cluster is in the cloud, **be sure to delete it when you no longer need it**. Failure to do this will incur unwanted costs and consume unnecessary energy and resources.

Chapter summary

In this chapter, you learned how to use a Kubernetes Deployment to perform a rolling update.

You edited the Deployment YAML file and added instructions to control the flow of the update. You also updated the version of the application image and sent the updated configuration to Kubernetes. Finally, you monitored and verified the operation.

10: What next

Congratulations on finishing the book, I hope you loved it!

If you read it all and followed the examples, you've learned the basics, and you're ready for your next steps.

Here are some quick suggestions for next steps. And yes, I'm recommending a bunch of my own stuff. But here's the truth.

- If you like this book, you'll love my other stuff
- I'm super busy and don't get a chance to read and test other people's stuff

Of course, if you didn't like this, I'm gutted. But that's life, and you probably won't like my other stuff either. If that's you, I'd love you to email me at **qsk-book@nigelpoulton.com** and tell me what you didn't like.

Books

If you like books and want to continue your Kubernetes journey, check out **The Kubernetes Book**. It follows on from here, goes into a lot more detail, is regularly listed as a best-seller on Amazon, and has the most Amazon ratings and reviews of any book on Kubernetes. It's also updated annually.

If you liked this book and want a similar introduction to Docker, check out my **Getting Started with Docker** book.

If you're into certifications, **The KCNA Book** is designed to help you smash the *Kubernetes and Cloud Native Associate* exam. It covers all exam topics and includes all of the following:

- Extensive explanations and examples
- Flashcard-style recaps
- Over 250 review questions

- A complete 60-question sample exam

None of the questions are from the actual exam. However, they are all written in the same style as exam questions. If you want your first Kubernetes certification, this book will get you there!

Many of my books are also available in audio format, so you can learn on the go.

If you're unsure about technical books in audio format, the following Audible reviews should help.

Video courses

If you like video courses, I've got lots on pluralsight.com. They're a lot of fun and apparently *"laugh out-loud funny"* — not my words.

Events

I'm a huge fan of community events.

My favourite in-person event is KubeCon, and I recommend you attend if you can. You'll meet great people and learn a lot from the sessions.

I also recommend local community meetups. Just google any of the following to find one local to you. You'll need to temporarily disable any VPN or other browser privacy tools for those searches to work ;-)

- "Kubernetes meetup near me"
- "Cloud native meetup near me"

Show some love

I'd consider it a personal favor if you write a short review or give the book some stars on Amazon. You can leave an Amazon review if you got the book from somewhere else. Cheers!

Let's connect

Finally, thanks again for reading my book. Feel free to connect with me on any of the usual platforms, and we can discuss Kubernetes and other cool technologies.

- linkedin.com/in/nigelpoulton
- @nigelpoulton@hachyderm.io
- @nigelpoulton
- @nigelpoulton.bsky.social
- gsd@nigelpoulton.com

Appendix A: Lab code

This appendix contains all the lab exercises from the book, in order. It assumes you've got a Kubernetes cluster, installed Docker, installed Git, and configured `kubectl` to talk to your cluster.

Chapter 5: Creating a containerized app

Clone the book's GitHub repo.

```
$ git clone https://github.com/nigelpoulton/qsk-book.git
Cloning into 'qsk-book'...
```

Change into the **qsk-book/App** directory and run an `ls` command to list its contents.

```
$ cd qsk-book/App

$ ls
Dockerfile   app.js   bootstrap.css
package.json   views
```

Run the following command to build the application into a container image. Be sure to run it from within the **App** directory. If you have a Docker Hub account, make sure you use your own Docker account ID.

```
$ docker image build -t nigelpoulton/qsk-book:1.0 .

[+] Building 66.9s (7/7) FINISHED                        0.1s
<Snip>
=> naming to docker.io/nigelpoulton/qsk-book:1.0         0.0s
```

Verify the image was created and is present on your local machine.

```
$ docker image ls
REPOSITORY             TAG    IMAGE ID       CREATED          SIZE
nigelpoulton/qsk-book  1.0    c5f4c6f43da5   19 seconds ago   84MB
```

Push the image to Docker Hub. This step will only work if you have a Docker account. Remember to substitute your Docker account ID.

```
$ docker image push nigelpoulton/qsk-book:1.0

The push refers to repository [docker.io/nigelpoulton/qsk-book]
05a49feb9814: Pushed
66443c37f4d4: Pushed
101dc6329845: Pushed
dc8a57695d7b: Pushed
7466fca84fd0: Pushed
<Snip>
c5f4c6f43da5: Pushed
1.0: digest: sha256:c5f4c6f43da5f0a...aee06961408 size: 1787
```

Chapter 6: Running an app on Kubernetes

List the Nodes in your K8s cluster.

```
$ kubectl get nodes
NAME             STATUS   ROLES    AGE   VERSION
k3s-qsk...d5idm  Ready    <none>   5m    v1.28.2+k3s1
k3s-qsk...jhnpn  Ready    <none>   5m    v1.28.2+k3s1
k3s-qsk...um2r2  Ready    <none>   5m    v1.28.2+k3s1
```

Run the following from the root of the GitHub repo. If you're currently in the **App** directory, you'll need to run the `cd ..` command to back up one level.

Deploy the application defined in **pod.yml**.

```
$ kubectl apply -f pod.yml
pod/first-pod created
```

Check the Pod is running.

```
$ kubectl get pods
NAME        READY   STATUS    RESTARTS   AGE
first-pod   1/1     Running   0          10s
```

Get detailed info about the running Pod. The output has been snipped.

```
$ kubectl describe pod first-pod

Name:         first-pod
Namespace:    default
Node:         k3s-qsk-885e/192.168.1.4
Labels:       project=qsk-book
Status:       Running
IP:           10.42.2.3
<Snip>
```

Deploy the Service.

```
$ kubectl apply -f svc.yml'
service/svc-lb created
```

Check the external IP (public IP) of the Service. Your Service will only have an external IP if it's running on a cloud.

```
$ kubectl get svc
NAME     TYPE           CLUSTER-IP    EXTERNAL-IP    PORT(S)
svc-lb   LoadBalancer   10.43.63.98   74.220.22.94   5555:31384/TCP
```

Point your browser to the IP from the `EXTERAL-IP` column.

Run the following commands to delete the Pod and Service.

```
$ kubectl delete pod first-pod
pod "first-pod" deleted

$ kubectl delete svc svc-lb
service "svc-lb" deleted
```

Chapter 7: Adding self-healing

Run the following command to deploy the application specified in **deploy.yml**. This will deploy the app with five Pod replicas.

```
$ kubectl apply -f deploy.yml
deployment.apps/qsk-deploy created
```

Check the status of the Deployment and Pods it is managing.

```
$ kubectl get deployments
NAME         READY   UP-TO-DATE   AVAILABLE   AGE
qsk-deploy   5/5     5            5           14s

$ kubectl get pods
NAME                     READY   STATUS    RESTARTS   AGE
NAME                     READY   STATUS    RESTARTS   AGE
qsk-deploy-84...5txzv    1/1     Running   0          31s
qsk-deploy-84...mbscc    1/1     Running   0          31s
qsk-deploy-84...mr4d8    1/1     Running   0          31s
qsk-deploy-84...nwr6z    1/1     Running   0          31s
qsk-deploy-84...whsnt    1/1     Running   0          31s
```

Delete one of the Pods. Be sure to use the name of a Pod from your environment.

```
$ kubectl delete pod qsk-deploy-845b58bd85-5txzv
pod "qsk-deploy-845b58bd85-5txzv" deleted
```

List the Pods to see the new Pod Kubernetes automatically started.

```
$ kubectl get pods
NAME                          READY   STATUS    RESTARTS   AGE
qsk-deploy-845b58bd85-2dx4s   1/1     Running   0          35s
qsk-deploy-845b58bd85-mbscc   1/1     Running   0          2m51s
qsk-deploy-845b58bd85-mr4d8   1/1     Running   0          2m51s
qsk-deploy-845b58bd85-nwr6z   1/1     Running   0          2m51s
qsk-deploy-845b58bd85-whsnt   1/1     Running   0          2m51s
```

The new Pod is the one that's been running for less time than the others.

Chapter 8: Scaling an app

Edit the **deploy.yml** file and change the number of replicas from five to ten. **Save your changes**.

Re-send the Deployment to Kubernetes.

```
$ kubectl apply -f deploy.yml
deployment.apps/qsk-deploy configured
```

Check the status of the Deployment.

```
$ kubectl get deployment qsk-deploy
NAME         READY   UP-TO-DATE   AVAILABLE   AGE
qsk-deploy   10/10   10           10          2m29s
```

Scale the app down with `kubectl scale`.

```
$ kubectl scale --replicas 5 deployment/qsk-deploy
deployment.apps/qsk-deploy scaled
```

Check the number of Pods.

```
$ kubectl get pods
NAME                           READY   STATUS    RESTARTS   AGE
qsk-deploy-845b58bd85-2dx4s    1/1     Running   0          3m7s
qsk-deploy-845b58bd85-mbscc    1/1     Running   0          3m7s
qsk-deploy-845b58bd85-mr4d8    1/1     Running   0          3m7s
qsk-deploy-845b58bd85-nwr6z    1/1     Running   0          3m7s
qsk-deploy-845b58bd85-whsnt    1/1     Running   0          3m7s
```

Edit the **deploy.yml** file and set the number of replicas back to five and **save your changes**.

Chapter 9: Performing a rolling update

Edit the **deploy.yml** file and change the image version from **1.0** to **1.1**.

Add the following lines in the **spec** section. See **rolling-update.yml** for reference.

```
minReadySeconds: 20
strategy:
  type: RollingUpdate
  rollingUpdate:
    maxSurge: 1
    maxUnavailable: 0
```

Save your changes.

Send the updated YAML file to Kubernetes.

```
$ kubectl apply -f deploy.yml
deployment.apps/qsk-deploy configured
```

Check the status of the rolling update.

```
$ kubectl rollout status deployment qsk-deploy
Waiting to finish: 1 out of 5 new replicas have been updated...
Waiting to finish: 1 out of 5 new replicas have been updated...
Waiting to finish: 2 out of 5 new replicas have been updated...
<Snip>
```

The following commands will clean up by deleting the Deployment and Service objects.

```
$ kubectl delete deployment qsk-deploy
deployment.apps "qsk-deploy" deleted

$ kubectl delete svc svc-lb
service "svc-lb" deleted
```

If your Kubernetes cluster is running in the cloud, remember to delete it when you're done. This avoids wasting resources, energy, and money.

Terminology

This glossary defines some of the most common Kubernetes-related terms used in the book. I've only included terms used in the book. For a more comprehensive coverage of Kubernetes, see ***The Kubernetes Book***.

Ping me if you think I've missed anything important:

- qskbook@nigelpoulton.com
- https://nigelpoulton.com/contact-us
- https://twitter.com/nigelpoulton
- https://www.linkedin.com/in/nigelpoulton/

As always, I know that some of you are passionate about definitions of technical terms. That's OK, and I'm not saying my definitions are better than anyone else's — they're just here to be helpful.

Term	Definition (according to Nigel)
API Server	Part of the Kubernetes control plane and runs on all control-plane nodes. All communication with Kubernetes goes through the API Server. **kubectl** commands and responses go through the API Server.
Container	Application and dependencies packaged to run on Docker or Kubernetes. As well as application stuff, every container is an isolated *virtual operating system* with its own process tree, filesystem, shared memory, and more.
Cloud-native	An application that can self-heal, scale on-demand, and can perform rolling updates and rollbacks. They're usually microservices apps and run on Kubernetes.

Term	Definition (according to Nigel)
Container runtime	Low-level software running on every Kubernetes worker node. Responsible for pulling container images and starting and stopping containers. The most famous container runtime is Docker, however, **containerd** is now the most popular container runtime used by Kubernetes.
Controller	Control plane process running as a reconciliation loop monitoring the cluster and ensuring the observed state of the cluster matches desired state.
Control plane node	Cluster node running control plane services. The brains of a Kubernetes cluster. You should deploy three or five for high availability.
Cluster store	Part of the control plane that holds the state of the cluster and apps.
Deployment	Controller that deploys and manages a set of stateless Pods. Performs rolling updates and rollbacks and can self-heal from Pod failures.
Desired state	What the cluster and apps should be like. For example, an application's *desired state* might be five replicas of xyz container listening on port 8080/tcp.
K8s	Shorthand way to write Kubernetes. The "8" replaces the eight characters in *Kubernetes* between the "K" and the "s". Pronounced "Kates".
kubectl	Kubernetes command line tool. Sends commands to the API Server and queries state via the API Server.
Kubelet	The main Kubernetes agent running on every cluster node. It watches the API Server for new work assignments and maintains a reporting channel back.

Term	Definition (according to Nigel)
Label	Metadata applied to objects for grouping. For example, Services send traffic to Pods based on matching labels.
Manifest file	YAML file that holds the configuration of one or more Kubernetes objects. For example, a Service manifest file is typically a YAML file that holds the configuration of a Service object. When you post a manifest file to the API Server, its configuration is deployed to the cluster.
Microservices	A design pattern for modern applications. Application features are broken into their own small applications (microservices/containers) and communicate via APIs. They work together to form a useful application.
Node	Also known as worker node. The nodes in a cluster that run user applications. Must run the kubelet process and a container runtime.
Observed state	Also known as *current state* or *actual state*. The most up-to-date view of the cluster and running applications.
Orchestrator	Software that deploys and manages microservices apps. Kubernetes is the most popular orchestrator of microservices apps.
Pod	A thin wrapper that enables containers to run on Kubernetes. Defined in a YAML file. The smallest unit of deployment on a Kubernetes cluster.
Reconciliation loop	A controller process watching the state of the cluster via the API Server, ensuring observed state matches desired state. The Deployment controller runs as a reconciliation loop.

Term	Definition (according to Nigel)
Service	Capital "S". Kubernetes object for providing network access to apps running in Pods. Can integrate with cloud platforms and provision internet-facing load balancers.
YAML	Yet Another Markup Language. Kubernetes configuration files are written in YAML.

Index

I

K

L

M

N

O

P

R

S

U

W

Y

More from the author

#1 New Release ★★★★★ 3 ratings

Brand new in October 2023, **Getting Started with Docker** is the best, and fastest way to learn Docker and containers. Whether you're a developer, sysadmin, architect, management, or even sales, the easy-to-follow examples in this book will get you up-to-speed in no time.

DOCKER
DEEP DIVE
ZERO TO DOCKER IN A SINGLE BOOK
2023 Edition
By Docker Captain
Nigel Poulton

Show some love!

Enjoyed the book!

Head over to Amazon and give it some stars and a review.

Writing books is incredibly hard, and I spent many late nights and early mornings making this book as amazing as possible for you. Taking a couple of minutes to leave a review would be great.
Thanks!

Made in the USA
Las Vegas, NV
20 April 2024